A Wild Neighborhood

To Harold and Leona with greetings from the Wild Neighborhood!

John Henricksson

The University of Minnesota Press
gratefully acknowledges assistance provided
for the publication of this volume by the
John K. and Elsie Lampert Fesler Fund.

A WILD Neighborhood

John Henricksson

Illustrated by Betsy Bowen

University of Minnesota Press

Minneapolis • London

Portions of chapter 7 are printed from "Tales of Quetico-Superior Area, in conjunction with J. W. White's *Historical Sketches of the Quetico-Superior Area,"* by Peter J. Nowak. Paper submitted to Dr. Julius Wolff, University of Minnesota, Duluth, 1972.

Published by the University of Minnesota Press
111 Third Avenue South, Suite 290
Minneapolis, MN 55401-2520

http://www.upress.umn.edu

Printed in the United States of America on acid-free paper

Library of Congress Cataloging-in-Publication Data

Henricksson, John.
A wild neighborhood / John Henricksson
p. cm.
Includes bibliographical references (p.).
ISBN 0-8166-3017-8 (pb : alk. paper)
1. Zoology—Minnesota—Gunflint Lake Region. I. Title
QL84.22.M6H45 1997
591.9776'75—dc21 97-11452

For Anna and Eva;
may they someday take to the woods.

It seems that there both men and wild beasts pursued their own paths freely and, as if conscious of the wild freedom of their world, molested one another not at all.

—Rockwell Kent,
A Journey of Quiet Adventure in Alaska

Contents

Preface xi

Acknowledgments xix

1
Most like a Man 1

2
First Citizen of Minnesota 17

3
Hunter of the High Places 31

4
A Parliament of Owls 43

5
The Magnificent Seven-Footer 57

6
Rascal in a Gray Suit 71

7
The Empire Builder 83

CONTENTS

8
Along Came a Spider 99

9
Helen Hoover's Deer 109

10
More than a Bird 119

11
Icon of the Wilderness 133

12
Ghosts 145

Recommended Reading 163

Preface

One crisp October morning during my daily wood-chopping hour at our Gunflint Lake cabin I glanced up the driveway and was delighted to see I had a visitor, an adult black bear sitting spraddle-legged in the gravel, its front paws folded serenely over its generous belly, watching me work. Pretending to be oblivious to this shaggy spectator, I continued splitting and stacking, watching him warily out of the corner of my eye. After about ten minutes he got to his feet, scratched all over, yawned mightily, and lumbered off into the woods. Perhaps the bear was curious, perhaps tired, or maybe just bored, but I was enriched by his visit, and aware that such an experience was as intimate as any with a wild creature should ever be.

That bear was probably one of the neighbors who used to visit occasionally as a cub. He was about two years old then and just being sent out into the world by the mother bear to find his own territory. He would wander across the cabin deck in the evening and early morning checking out the bird feeders. Even the dog got so used to his visits she quit barking. The cub had a large white chest blaze then and a curiously mangled ear. Now, I noticed, the blaze is much smaller, but the ear still sticks out at an odd angle. His visits are less frequent, but when he shows up I am always reminded how fortunate we are to share this neighborhood with wild creatures.

The word *neighbor* has old dictionary meanings; "near dwellers" and "borderers" are two I particularly like. There is something intimate, but a little shadowy, about the words, much like the wild creatures in this narrow wedge of the Superior National Forest, tucked into the far northeastern corner of Minnesota between the North Shore of Lake Superior and the Boundary Waters canoe trails of the Minnesota-Ontario wilderness. It is not a true wilderness, as it is often wistfully, but incorrectly, called. Older residents call it bush country, a Canadian term meaning uncultivated land abandoned to natural growth.

I see the creatures only occasionally, but am always aware of their presence from tracks, squeaks, scat, claw marks, and wild music. Our relationship over the years seems based on a very proper distance and restraint, but they seem as unafraid and curious as I have become. When we do meet, there is sometimes a little tension, but never any panic or aggression, and, as in all neighborhoods, some residents are more accessible than others. The timber wolf, should I ever get within range of his superacute senses, vanishes instantly, but this black bear seems almost sociable at times, and the gray jay may even poke its head in my jacket pocket searching for a snack.

The neighborhood today is the boreal forest at its venerable best, old-growth patches of several thousand acres. Here the great white pines were mature when the Declaration of Independence was signed. The blocky, white cedars may be five hundred years old, and the paper birches, growing in the sunnier places, lost their glistening whiteness decades ago. Now, large pewter-colored shards of bark curl away from their massive trunks. We once counted the annual rings of a felled Goliath birch back to 1853; it was a sapling during the Civil War. The interlacing canopies of these old giants allow only puddles of sunlight to dapple the forest floor, which is a carpet of shade-tolerant plants such as bunchberry, coral root, clintonia, twisted stalk, and the fairyland feathermoss. Scattered through the woods stand solemn erratics, the huge boulders deposited by the last glacier, many weighing several tons. These irregular slabs of pink Saganaga granite, or perhaps of some far-off Arctic formation, were chiseled off and scooped up by the inexorable ice movement during the late Pleistocene epoch, until melting or shifting deposited them here. One monolith, standing in the forest near Magnetic Lake, is several stories high.

The lumpy, moss-covered shapes of decomposing tree trunks and upended roots called windthrows add a further ghostly presence in the half-light that pervades this old forest even at midday. Beneath the trees the land is billions of years old, the resting product of folding and faulting mountain ranges, volcanos, a lifeless tropical sea, a two-mile-thick blanket of ice, and now lakes and forests. The thin, acid soil covering the Canadian Shield bedrock is largely decomposing forest duff, some reddish clays, glacial drift, and silt. Some of the creatures here, like the pine marten, are indigenous to the old-growth forest. Others, like the deer, bear, and most perching birds, prefer the nearby mix of intermingled second-growth woods a few miles to the south—clear-cuts, roadsides, forest edges, bogs, and shorelines. Here the highbush cranberry, aspen, beaked hazel, mountain ash, and pin cherries provide a nourishing and varied menu for many wild palates.

It is at the southernmost boundary of the snowy owl's range, but near the northern limits for the little green heron. The long winters of deep snow and ice-locked lakes and rivers severely limit mobility and food supplies. Perhaps these shared hardships of winter make them more congenial, but creatures like pine martens, white-tailed deer, and gray jays seem to be always hanging out around the cabin during winter months. The nervous and wary timber wolves can sometimes be spotted across the lake on sunny days, and the elegant stitchery of the voles appears every morning after they have traced their nightly wanderings over the snowdrifts before descending near daylight into their world of tunnels, chambers, and food caches beneath the snow.

Early in December, one big snow year, a white-tailed buck deer with a perfectly symmetrical, eight-point rack began appearing at the cabin every morning for a bucket of corn I scattered for him. As his normal food sources became buried under the endless snow, it seemed harder and harder for him to get around, and he became very emaciated. Sharp hipbones stuck up high off his rump, and his snow-fence ribs heaved and protruded through his ragged coat as the winter progressed. But he kept going on the corn ration he picked up every morning near the woodshed. Later in the winter when the bucks normally cast their antlers, he laid his polished crown near the cabin where I could find it easily. Coincidence? Of course it was, but I chose to think it was an expression of gratitude.

Biologists scoff when they encounter anthropomorphism, the sentimental but mistaken attributing of human emotions to animal behavior. The logic of their science is overwhelming, but I am certain I saw gratitude in the deer's gift, just as I feel the wrath of a scolding red squirrel.

To the Anishinabe (Ojibwe), the original people of this region, all creatures, including man, were related and dependent upon one another. There was much communication among them, as well as a deep sense of spiritual oneness. The heritage of this appealing relationship has not disappeared completely, even from the subconscious of some scientists, who occasionally reveal their susceptibility to the beauty and intuition of the Indian legends.

Anthropologist Richard K. Nelson hints broadly at this departure from pure science in his ethnographic study of the Koyukon people of Alaska. In *Make Prayers to the Raven*, he tells how he became aware of a "rich and eloquent natural history unknown or ignored by my own culture." If we can reach out to this rich and eloquent natural history of the Indian people, perhaps we can extend some humanity to our wild neighbors without humanizing them. Observing them in fact and fancy is a good way to start.

Here is the black bear who dances on the weathered boards of my deck, leaving claw marks as beautiful as a craftsman's carving. Here, too, is the barred owl, the bird voyageurs called "le chat huant du nord," the hooting cat of the North. The ungainly but curiously dignified moose at the shoreline is the largest antlered mammal ever to roam the continent and seems smug in his role as lord of the forest. The cheeky bird up in the spruce tree that looks like it is wearing a tuxedo is the gray jay, the dandy of the crow family. Some call it the whiskey jack or camp robber, but it is the friendliest bird in the forest and often drops in for breakfast or accompanies me on winter trails. There are many near dwellers and borderers here, both furred and feathered, and most of them tolerate me as long as I understand my place in their world.

Acknowledgments

A naturalist is a person with a comprehensive and eclectic knowledge of the natural world. He or she is often a botanist or a zoologist, but still a generalist, one with interests ranging across the whole spectrum of natural history.

Perhaps several steps below that more authoritative title would be the observer, or reporter, a label I feel more comfortable with. The natural neighborhood of the Gunflint region fascinates and delights me, and I need no more in the way of credentials because I am lucky enough to have all around me a sort of bush country Internet, a large group of qualified people with all of the necessary academic and practical knowledge I often search for.

Among those who have been especially helpful to me during the preparation of these essays on my wild neighbors are Department of Natural Resources (DNR) area wildlife manager Bill Peterson; DNR ecological biologist Steve Wilson; DNR research biologist Bill Berg; U.S. Forest Service ecologist Wayne Russ; and ornithologists Laura Erickson, Bill Lane, and Kim

Eckert. There are many neighbors, perhaps without professional qualifications, but with years of experience and wildlife knowledge, whose help was invaluable: Jon and Mary Ofjord, Ken and Molly Hoffman, Frank and Pat Shunn, and Our Lady of the Trail, Justine Kerfoot.

Special thanks are due my wife Julie, who shares the trails with me and untangles the mysteries of syntax, grammar, and the intimidating computer. And finally, to a friend long gone, Charlie Ott, who taught me a special reverence for the Gunflint region.

1
Most like a Man

EVEN THOUGH WE have no theaters, shopping malls, or video stores in the wild neighborhood, there is plenty of entertainment, and we don't feel deprived. Quite the opposite. The natural world of this boreal forest region provides grand theater, and the stage is constantly busy, changing scenery, playbills, and casts with the seasons. The timber wolves provide drama, and moose and bald eagles bring majesty to the stage, with a musical score by a variety of warblers, the olive-sided thrush, and, of course, the loons. We get mystery from the owls, joy and pathos from the deer, bawdy comment from the ravens, ongoing criticism of everything from the red squirrels, and, often, comedy from the bears.

Bears and their antics dominate most of the storytelling here, probably because of their boldness and the audacity of their escapades. Nate Rusk, up on Seagull Lake, confronts a large, shaggy visitor in her kitchen who is eyeing a freshly baked blueberry pie cooling on the ledge. Greg Gecas shoos a pair of cubs away from one of his cabins, both of them guilty of breaking and entering, then stealing a pound of bacon from the refrigerator. The Tiffanys found their grandchildren nose-to-nose with a curious bear, separated only by the window glass in their living room. There was even a cholesterol-free bear around Devil Track Lake who regularly selected only oat bran cereal from its favorite cabin pantry.

The Becklunds, down near town, could sometimes watch as many as seven bears from their kitchen window. One regular visitor, a matronly female, took to lounging around on their porch furniture. One time they decided enough was enough, so they filled a lot of balloons with ammonia gas, smeared them with honey, and placed them in a row near their garden. The idea was that the rough tongue and sharp teeth of the bears would puncture the balloons and give them a faceful of the foul-smelling gas—a homemade mace attack. The bears thought it was a great idea. They licked all the balloons clean and never broke one. Jack finally resorted to a wrist-rocket slingshot and a couple of loud shepherd dogs. That worked better, but the lounger continued to use the deck chair.

For a time, we too had a regular visitor at the cabin—a second-year, shiny black cub who would stand up on his tiptoes to reach the hanging dome bird feeder on our front porch. By stretching to his full height, he would take a mighty swing at the feeder, barely ticking the bottom so it would swing, pouring sunflower seeds all over the floor. When he had a goodly supply piled up down there, he would lie down on his belly and lick up all the seeds. The next year he was big enough so this technique was unnecessary. He would just stand up on his hind legs, hug the feeder to his chest and sag downward, straightening the hook, and trot off into the woods with his prize. He never went far. I would follow at a safe distance a little later and retrieve the empty feeder, chewed and scratched, but still serviceable.

We have an ambivalent relationship with the American black bear (*Ursus americanus*), an original resident of the old-growth forest. The bear is Muckwa to the Ojibwe, a totemic symbol of strength and courage and one of the five god creatures who rose from the sea to create the earth at the very beginning. The ambivalence is composed of respect, fear, curiosity, awe, affection, and a certain sense of familiarity. This is the creature Chief Dan George portrayed as being "most like a man."

There are many reasons for the perceived relationship between man and the bear. Some have a spiritual cause, and others are due to physical similarities. The bear was "elder brother" to the Anishinabe. According to J. G. Kohl in *Kitchi-Gami,* "Of all the animals in the forest, the Indians respect the bear most. They regard it almost in the light of a human being. Indeed, they will often say that the bear is an Anishinabe. They will converse with it, thinking all the while the bear understands them. . . . It may be easily understood how the Indians see an enchanted being in them."

In their seminal work on the bear in nature, myth, and literature, *The Sacred Paw,* Paul Shepard and Barry Sanders describe that similarity:

> Like us, the bear stands upright on the soles of his feet, his eyes nearly on a frontal plane. The bear moves his forelimbs freely in their shoulder sockets, sits on his tail end, one leg folded like an adolescent slouched at the table, worries with moans and sighs, courts with demonstratable affection, . . . snores in his sleep, spanks his children, is avid for sweets and has a moody, gruff, morose side. . . . [He is] wily, strong, agile; independent in ways that we humans left behind when we took up residence in the city. He is an ideogram of man in the wilderness.

In spite of this perceived relationship and our feelings of affection for this magnificent animal, we must never feel too chummy. The fundamental rules still apply: This is an immensely strong, fast, and wild creature with fearsome weapons, unpredictable and temperamental. We must never approach bears. Never even look them in the eye; this is intimidating and challenging to wild creatures.

Weights vary with habitat, but here the black bear can reach 400 pounds, although 250 to 300 pounds is an average adult weight. It has a coat ranging from shades of cinnamon and mahogany to dark chocolate brown and jet black. Often the younger bears have white chest blazes, which disappear after three or four years. It is a foraging omnivore, which means that it wanders constantly and will eat anything that isn't freshly painted or bolted down. It is this nonselective and ravenous appetite that gets the black bear in so much trouble, often earning it the doomsday label of "nuisance bear." When, after many complaints, a bear is designated a nuisance, the authorities give permission to shoot it, or they try to relocate it by live trapping and carting it off to a new area, hoping that it will find a new home to its liking. More often, the bear will head right back to its old territory and resume its bad habits. Last year, there were many complaints about a young male entering cabins across the lake, stealing food and scaring the bejesus out of tourists. It was live trapped and taken down to the Mink Lake country about fifty miles south of here. When released, the bear shook itself off, got its bearings, and took off straight through the woods, arriving back at Gunflint in about twenty-four hours. In a day or so, it was shot by a resorter who feared the worst after seeing the bear snuffling around some of the guest cabins.

Such an occasion is a sad one for those of us who enjoy having the bears around, but, like most of the injustices suffered by wild creatures, it is basically a man-made problem. The black bear is a curious and bold creature with a permanent case of the gut-rumbles, and this often leads it into intolerable situations in areas like this one, with its attraction for tourists. Years ago, the local governments designated areas to be bulldozed out, fenced, and used as garbage dumps; all of them immediately became open-air fast food cafés for the bears, and great tourist attractions, as well. When boredom set in at the cabin, someone was bound to say, "Let's go to the dump and watch the bears." Cars filled with bear watchers would line the perimeter of the dumps.

This situation lasted for years, but then environmental laws and some close calls with people who tried to feed the bears by hand forced the closing of the dumps in favor of recycling centers and sanitary landfills. This had many advantages, but it left the bears with a serious addiction problem and no free lunch counter. Soon they were back rattling around resorts, cabins, campgrounds, and roadside ditches. The bear always subordinates property rights to its growling stomach and becomes a persistent and unwelcome trespasser whenever its sensitive nose picks up the tantalizing aroma of garbage, or any kind of food. Many hunters have turned this character weakness of the black bears to their advantage, all with the questionable approval of the Department of Natural Resources.

The preferred method of bear hunting in this region is by baiting, that is, hauling garbage out into the woods to a preselected spot, letting it ripen to attract bears, and then, when the season opens, sitting in a blind until one comes along and shooting it.

"No fair," shout the antibaiting activists. "That's just like shooting a grazing cow."

"Sure as hell beats walking through the woods looking for 'em," says a hunter neighbor.

Some eat the meat, but to the hunter, the bear is primarily a trophy. To the Ojibwe, the bear was sustenance—food for survival, fur for blanket warmth, bones for utensils, claws and teeth for ceremonial adornment. According to Basil Johnston in *Ojibwe Hertiage,* when an Ojibwe hunter killed a bear, he made gifts to its spirit and prayed:

I had need,
I have dispossessed you of beauty, grace, and life.
I have sundered your spirit from its worldly frame.
No more will you run in freedom
Because of my need.
And I shall not want.

What a long journey it has been for Muckwa, the black bear, from the pantheon of the gods to the garbage pile.

The black bear is pretty laid back and slow moving, which probably accounts in part for its longevity. The oldest one from the wild was recorded at thirty-two years. Hunting pressure brings the average way down, but ten-year-olds are not uncommon in the more remote regions. Generally, the bear moves at a leisurely pace, sniffing, snoozing and scratching its way through the woods, although it can attain great speeds when it is charging prey or defending a cub. It grows slowly, spending five to seven months in a semicomatose hibernation in the den, where the cubs are born after a very long gestation period—about 215 days. The bear is one of the prochoice advocates among wild creatures. Mating takes place in the late spring, and if the berry crop and other succulents they load up on before hibernating fail or are too sparse, the pregnant female will self-abort the fertilized ova, which are floating in the uterus until she instinctively makes the decision whether or not to carry the cubs to full term. This process is still only vaguely understood but is related to body weight and other physical conditions.

Mother black bear is an excellent parent—protective and strict. She has cubs every other year, so she keeps them with her for two years before sending them out on their own.

One oppressively hot, still August morning I pushed the canoe into Dog Ear Bay off Gunflint's south shore. The bay is a long, pointy wedge of water with a round lobe at one end, peppered with boulders in the middle and fringed with sweet gale and spiky reed grass. Down at the shallow, mucky end where a sometimes creek wanders in, bog rosemary and Labrador tea grow thick in a tangle just right for nesting loons. The length of the south shore is rimmed by heavily forested palisades rising two hundred feet into the hazy morning mist. Farther down the shore, there is a grove of young aspens, straight, gray-green picket poles frothed with a topping of the palest green. With an occasional dip of the paddle, the canoe slid over tranquil water, its shadow undulating like a cruising dragon over the cobbled bottom. I rested the paddle over my knees and watched that cloud of aspen foliage seeming to float over a rock-strewn landscape.

Suddenly, I became aware of a movement I hadn't seen before. There was a patch of leaves in motion near the middle of the leaf cloud—a truly quaking aspen. For a moment all would be perfectly still, then quite suddenly, a violent tremor in the leaves would ripple outward like a plunging stone in a quiet pool. It would start, sway violently, then stop again, and all would be still. After this had gone on for several minutes, I decided to investigate.

I guided the canoe slowly next to a fallen log pointing out from shore, stepped out into the shallow water, and waded through the tall grass, which would conceal me for the moment. I wanted to approach the grove from above, thinking I'd have a better look into it if I were higher, so I went up the hill as carefully and quietly as I could, stooping low and avoiding the grasp of the gnarled alder bushes that thinned as I got farther up the slope. Up there were big granite slabs and stout cedars to hide behind.

Moving slowly, continuing to crouch low, I worked my way to a spot just above the aspens where a spruce top, snapped off by the wind, leaned down against the trunk, still connected by a wide strip of bark, making a perfect blind. I scuttled in among the fragrant boughs, pushing aside some small branches to make an opening big enough to see through.

In a few moments, I spotted the cause of the commotion. At the base of one of the supple young trees was a bear cub, about the size of a springer spaniel, furiously scratching its scrawny little butt against the rough bark of the tree. It would scratch and scratch, stop for a while, looking very relieved, and then start again, the top of the sapling waving crazily in the still morning air.

I watched, fascinated, for several minutes, never noticing the change in direction of a slight breeze that had come up since I first went ashore. Suddenly, from down the other side of the small hill came a throaty barklike cough: "Chuff, chuff . . . chuff, chuff." The cub went up on its hind legs, little round ears swiveling toward the sound, and took off at top speed down the hill and along the shore. Mother bear had called from the tangled woods to the east end of the bay, and Junior knew better than not to obey immediately. I began to back out of the tattered spruce when a thought occurred to me: how long had she been there watching me? I'll never know. It was almost as though she had let me admire her cub for a while, but how much time she would allow for this was her decision.

We fret a lot about the future of some of our wild neighbors. Man's intervention into the natural succession of the forest accelerates the whole process, and agriculture, logging, pollution, development, and mining take a heavy toll on the habitat. Some creatures seem to be more sensitive than others to these disruptions, but I think Muckwa, the black bear, will be here for a long time. It is a survivor like man, clever and adaptable. This Anishinabe symbol from creation, this "animal most like a man," may owe its survival to its mysterious relationship with humankind.

2
First Citizen of Minnesota

ON ONE OF EARLY June's more hospitable mornings, Julie and I were making our desultory way along the bog shoreline of Gunflint's narrow, westmost bay, canoeing with no destination, just poking around. The little bay was calm, but out on the big lake little gusts of wind sent wavelets skittering around in several directions, as though the morning wind hadn't yet made up its mind about a direction.

Early June in the north is seldom the best time for comfortable canoeing. It can be cantankerous, with an occasional cold squall dragging along in the wake of retreating spring, savage hordes of mosquitoes, and, once in a while, a misplaced November dawn that rimes the edges of paddles. But it is also a good time. Good for nesting loons, walleye fishing, and the glory days of the moccasin flowers, so we take a chance hoping for one of the surprises wilderness mornings can sometimes deal the gambler.

We spend quite a bit of time looning in the spring, because the normally super-shy loon gets a little careless during its preoccupation with romance and does not seem to mind a canoe floating quietly nearby. The male swims in tight circles around the female, flaps his wings, and rocks back and forth. His lady watches demurely, but if she decides in his favor, they dance off together, frothing the water and making an awful fuss. They nest close to the water and aren't very clever about camouflaging it. It was a nest we were looking for that morning.

Near the end of the bay where an inconsistent creek wanders into the lake, the bank was shrubby with sweet gale. Gnarled beaked hazel grew into a webwork thicket typical of the northern bog. The soon-to-be-earth of decaying reeds was visible only occasionally where a narrow break in the dense foliage revealed the slickery mud of a beaver run.

Some careless canoe handling bumped us along the tussocky shoreline grass, and, in the process of pushing at the grass with the paddles to get the canoe back into deeper water, we were abruptly confronted by an irate loon, flame-eyed and very upset at our intrusion. We assumed it was the mother loon on the nest, but couldn't be sure because both sexes incubate the eggs and appear identical. It was probably Ms. Loon, because usually the female will cover the nest during daylight hours and the male takes over at night.

Presumably, he is more capable of handling nocturnal predators such as foxes and skunks. These loons had assembled (*built* is too strong a word for a loon's nest) a mat of reeds, sticks, and mud at the water's edge, which would allow the young to slide right out of bed into the water. They can swim before they can waddle. This beautiful harridan was lying flat on her belly, wings spread wide over the eggs, cursing and darting her dagger beak at the canoe paddle.

That could have been a dicey moment if the creature had been a mink, who will make a frenzied, slashing charge at anything, regardless of size. Even a red-winged blackbird will attack against the odds if provoked, but not the loon. We were being threatened by nature's grand poseur, bluffed by an almost weaponless creature. The loon is impotent as a threat largely because it is virtually helpless on land. Its landing gear is set so far back it is unable to balance itself or walk on land. It is a missile-bird in the water but very awkward on land. Loons do not "fly" underwater; that is, they don't use their wings for propulsion like some sea birds. The shearwater, the eiders, and the puffins are among the best of these underwater fliers. When the loon swims underwater, its wings are folded tightly against its body. Its propulsion comes from heavily muscled legs and its paddle-shaped feet.

It seems like heresy to say anything but shining words about the loon, icon of wilderness waters and First Citizen of Minnesota. This large black-and-white bird with the charming necklace and blood-red eyes who pokes around near our dock for crayfish, who drops from the April sky looking for the very first patch of open water; this bird whose manic signature to the day echoes through the minds of all who hear it; this cherished neighbor has become a star.

Its records top the charts. Its image, in varying degrees of quality, appears on T-shirts, bumper stickers, wind socks, playing cards, ashtrays, welcome mats, billboards, planters, beer, and license plates. It has become an industry and an object of worship.

It inspires the most purple of prose. One stricken soul described its necklace as "a pearly luminescence woven in black velvet." Its head is a lustrous obsidian, coal black until the sun brings out greenish lights. Its black-and-white checkered cloak is distinctive. According to an Inuit legend, it was a gift from Raven, the Creator, who tattooed the design on the loon's back with his wingtip dipped in charcoal dust. But many birds are as handsomely turned out as the loon—the king eider, the ring-necked pheasant, the snowy owl, and the wood duck, to name just a few. No, it has nothing to do with the loon's appearance. It's The Voice.

It sounds like a soul in torment, calling from the other side; a cry from distant, unseen places. Sigurd Olson said, "One lone call seemed to embody all the misery and tragedy in the world." In the world of north country nature, only the timber wolf can match the loon's eerie cadences. Only the cry of the loon has the power to haunt, to burrow down into the deep recesses of memory and return at strange and distant times. Many authors of loon books introduce them by telling the story of their first experience with The Voice and how it inspired them to write. Others may recall a camping trip or other wilderness experience, and years later discover that The Voice was their most vivid and lasting memory. Scientists may break it down into boring audiographs and sonograms, analyzing its pitch and structure but completely missing the message of lonely freedom it sends.

Only the common loon (*Gavia immer*) casts this spell. Its cousins are way outclassed. The red-throated loon quacks like a duck, and the best the Arctic loon can manage is an asthmatic whistle. Ours is the coffee table loon, the loon in our hearts, the loon of our pride. But there is a dark side of the loon; some of its behaviors don't square with its image.

Like all carnivorous predators it is a killing machine. Its great speed underwater and deep dives of one hundred feet or more make it a super-efficient hunter of fish and crustaceans.

It has another habit that may be a little disturbing to its worshipers: it has been observed drowning merganser chicks. The common merganser, a fish-eating, diving duck, hatches a clutch of nine to twelve chicks in late May. Some seem to be day-care providers and will adopt other chicks. We have seen flotillas of eighteen little ones trailing after a single hen. The merganser (the sawbill) has a long, tubular, serrated bill that is ideally adapted to catching fish, and this makes it a competitor to the loon and its chicks for food. When this competition threatens its food supply, the loon goes into its U-boat mode. With its swim bladder slightly deflated, it stays offshore, out from the merganser fleet, its head barely visible above the water and its bill angled upward like a periscope. When it zeros in on the group of merganser chicks feeding in the rocks near shore, it submerges, kicks forward, and speeds underwater toward its target. It grabs a chick's feet and holds it down until it drowns. Not a very neighborly thing to do, but protecting its food supply is critical, and the laws of the wild are immutable.

The common loon is nature's ultimate showman, an avian Liberace, who knows how to press all our buttons. It sings, it dances, it preens and flounces around on its beautiful stage in its gorgeous costume. But in the fall when most of its audience has left the theater and it has shed its checkered cloak for one of sooty gray during the molt, the old showman is seldom seen. It's almost as though it is ashamed to be seen without its makeup. The loon even loses The Voice then, never wasting it on an empty theater. In its winter retreat along the Gulf of Mexico, the South Atlantic, or in Chesapeake Bay, the loon is seldom recognized because that haunting voice is never heard and the nondescript gray feathering gives it anonymity.

Then, when May comes once again to the north country, the postcard loon will be back here on stage, regal, haughty, and tuxedoed, admiring itself in its watery mirror, once more conscious of its star status. Like most accomplished actors, its timing is impeccable. It is only in front of the footlights and delivering its lines when the house is full during the tourist season. The north is the loon's stage, and we are its adoring audience.

The source of the word *loon* is not definite. The Old English word *lumme,* meaning a lummox, or clumsy person, is a possibility. Or it could be from the Norse *lomr,* which means a lame person. The species name, *immer,* is a Swedish variant of *ember,* meaning the color of blackened ashes, referring to the bird's molting color.

Loons return to Gunflint Lake on the day the ice goes out, usually late in April or early May. They need big water because a long runway is absolutely necessary for takeoff. They run on top of the water, wings pumping, until they can get airborne, usually needing up to a quarter of a mile to get into the air. Surrounding high timber is also an advantage so they can see predators coming from the air, such as hawks, eagles, and owls, which prey on the chicks.

The loons' chicks are probably the sorriest-looking creatures of the spring. They are scrawny, covered with wispy down the color of dirt, and comically awkward. They can't stand for the first week, but push themselves around on their bellies with oversized feet. Standing and walking is a major project, and for a while they fall over on their chins every time they try, but water is the magic elixir. With the nest built right next to the water, the chicks can slide in without walking, and when they do, they transform immediately. They are strong, graceful swimmers as soon as they hit the water, and can dive well in a few weeks. They stick very close to the adult, and when it senses danger, it deflates the swim bladder and sinks down in the water far enough so the loon chicks can clamber onto its back, where they ride out the problem high and dry. They grow and feather fast and in a few weeks begin to give hints of the strength and beauty that will come by midsummer.

Like all its wild neighbors, the loon has a niche in the food chain, which means it is prey as well as predator. Last year we witnessed part of the drama of this unchangeable protein cycle. The Hatfield cabin is on a narrow point at the entrance to Dog Ear Bay, an ideal wildlife observation location. The bay is one of the few secluded spots on Gunflint, a nine-mile-long, east-west oriented lake, wide open to the prevailing winds and normally very choppy. With its narrow entrance and surrounding ridges, the bay is usually a calm haven. Birds and waterfowl are abundant; moose, deer, and bears come down to drink; martens, otters, and beavers are often seen around the shore; and the raptors patrol the sky overhead, drawn by the bounty of food.

Across the entrance to the bay from Hatfield's point is a small cove rimmed with reddish gravel and the above-water beginnings of a long, bouldery reef that angles out into the lake, a fishy spot where loons often gather.

On a late July morning we pulled into Hatfield's dock, knowing that Mary would have the coffeepot

on and we could get updated on the doings of the wild neighbors. They had recently witnessed an extraordinary sight, the killing of a young loon by a bald eagle. A mother loon and her two chicks had been feeding along the reef for the past few days. The chicks were about the size of mallards, nearly half grown, with dark gray backs and whitish breasts, but still unpatterned, marking them as immature loons. They were close to shore among the big rocks, apparently feeding on crayfish or minnows, when a bald eagle soared over the treetops that crowned the granite headland standing at the east entrance to the bay.

Spotting the loons, the eagle flared and dived, its yellow talons clenched and its wickedly curved beak held out in front like a raised saber. The attack came so suddenly that there was no chance for escape, and the loon chick exploded in a storm of gray feathers as the lightning bolt struck.

The two remaining loons dived for cover, and the eagle, thrashing around in the water, powered its way to shore, dragging the young loon. It fed for about fifteen minutes, then flew away carrying the leftovers in its talons. Bill and Mary canoed over to the spot immediately and found nothing but a few feathers, a foot, and a battered head.

We came a little later, and the Hatfields were curious about why they had seen neither of the remaining loons since the attack. Julie and I decided to canoe the bay's shoreline just in case we might find the nest or some evidence of hiding loons. We pushed the canoe along the brush and mud of the west end where the creek comes in and then down the heavily wooded south shore under the cliffs. About halfway down we spotted a silvery gray mass bobbing in the water, tangled in the branches of a deadfall aspen. Julie held the canoe steady while I climbed out and crawled along the trunk to take a closer look. It was a young loon, very likely the twin of the one killed by the eagle. But how had it died? I examined the bird very closely; there wasn't a drop of blood, a wound, or any evidence of injury.

I talked later with several experienced local birders about the dead loon, but no one seemed to have an answer. I went up to Saganaga to talk with an old friend who had been observing loons for many years to see if he might shed any light on the puzzle.

"Probably the mother killed it," he said. "They do that, you know."

"But why would the mother kill it?" I asked. A wordless shrug of the shoulders was the only answer I got. The mystery is still unsolved.

3
Hunter of the High Places

OUR FEW ACRES OF the old forest rise gradually from the bouldered shore to a dirt road in back and are dominated by a couple dozen old giants, virgin white pines whose sun-reaching strength has pumped them high above the surrounding trees and spread their irregular branches into a shimmering dark green crown. These big pines are the signature trees of the Gunflint region and are sentinel posts for bald eagles, gossip fences for the lugubrious ravens, and prime hunting grounds for the luxuriantly furred lightning bolt of the forest, the pine marten (*Martes americana*), the only predator that hunts these high places.

Watching the martens at play, I am captivated by their silky beauty and flowing movements. When they get down to work, hunting red squirrels in the big pines, they are the quintessence of control and electric speed.

In the late summer, the white pine's branches are tipped with caramel-colored cones that contain nuts the red squirrels gather excitedly, stuffing their cheek pouches full and racing to the ground to store them in secret places. The marten watches all this busyness, flattened on a branch or hidden in a cluster of needles. It seems to be measuring distances and checking escape routes until it is ready to make its move. When it breaks from cover, the red squirrel is already a statistic—well, not always. Little Red has some nifty moves of its own. It makes right angle turns at top speed, spirals like a curling vine around long branches, and sometimes, in a panic, makes a free-fall leap into space. From heights of sixty feet or more, this is often a fatal maneuver, but occasionally the squirrel will get a claw into something on the way down, swing itself onto all fours, and streak off, leaving a string of curses trailing in the air. The more frequent scenario has the marten killing the red squirrel in the first ten feet of the chase. The marten never pursues when the squirrel does its skydiving act, but I have seen them fall, and the marten, like the cat, has the ability to turn over in the air and land on its feet.

I have never seen a marten eat its prey in the tree. Usually it carries it down to the ground and takes it to the den if it has kits, or finds concealment where it can eat under cover. It seems very nervous when it is out in the open, standing on its hind legs and running in place like a waiting jogger, its head and upper body

swiveling and feinting, watching the heights and the ground around it. A barred owl or a rough-legged hawk is sometimes watching and could score a rare double if the marten weren't vigilant.

One winter morning I looked out the bedroom window and there, clinging to a cedar tree next to the cabin, was a mature pine marten looking in at me. It was a beautiful, prime specimen, mahogany-glossed, sleek, and onyx-eyed, with a pale orange throat patch and silvery hairs lining its foxlike ears. I assumed it was searching for food, so I took out a strip of bacon and nailed it to the tree. As I opened the door, the marten shot up the tree and across the cabin roof, but I was certain the irresistible smell of bacon would bring it down again as soon as I ducked into the cabin. Like its smaller relatives, the mink and the weasel, the marten has a very high metabolism and must eat constantly. Before I got back into the cabin I noticed it high up in a scraggly old balsam draped over a branch, looking like the whole fox fur my aunt used to wear over her shoulders.

It was back in about ten minutes, climbing cautiously up the cedar to the free end of the bacon strip. It sank its teeth into the bacon and tugged downward with all the strength in its little three-pound body. It shook its head violently from side to side, braced its feet wide, and tugged some more. No good. The bacon would not tear off the nail. This was a big problem. It was completely exposed on that bare tree trunk, and this really bothered it. Between bouts of jerking on the bacon strip, it would run up the tree to some low, bushy branches and assess the situation, chirp excitedly and run back down to resume its tugging. Nothing worked. It finally abandoned all caution and started chewing at the bottom of the strip, tearing off chunks with its scalpel-sharp incisor teeth, gulping them down quickly and tearing off some more until it finally reached the nailhead and chewed around it for the last bite. Finishing the bacon, the marten took a look around, dropped to the ground, and scooted under the cabin.

Some of my neighbors see martens at their bird feeders, especially if they put out some raspberry jam, the marten's all-time favorite treat and a favorite trapper's bait. Usually the marten subsists on red squirrels, voles, mice, an occasional snowshoe hare, berries, and nuts—and bacon when possible. In spring there are bird's eggs, and in summer, earthworms, frogs, and toads. Rosehips are a fall delicacy.

I never saw that marten at the window again, but that's not unusual. They never stay in one place for long. Martens are loners, and they roam over a large territory. The male and female are together for only a short time during the mating season in the late summer, but after that both continue wandering in their unending quest for food.

Most wild creatures are very wary when with their young. One will occasionally break that rule, and when that happens, I have a tendency to anthropomorphize the reasons; that is, I find some human trait in their behavior.

Hiking down the Iron Lake Road one spring morning, I saw a low-slung, dark brown dragon coming toward me about one hundred feet down the road. At least that's what it looked like: close to the roadbed, undulating, and about six feet long. It disappeared down a shallow dip, and I stopped, waiting for it to come over the rise. When it reappeared I realized I was seeing a mother pine marten with three little ones strung out behind. My only choice was to remain perfectly still. Had I tried to jump into the woods, the motion would have scared them off. All wild creatures can detect movement very quickly, and sometimes remaining perfectly still can be quite rewarding, as it was this time. The marten family kept on coming. Maybe they thought I was a tree standing in the middle of the road. Who knows what they thought? They couldn't help but see me. Here they came, humping along, the young following right in the mother's footprints, all about two feet apart. All four went right past no more than a few feet away, without ever looking up at me. It seemed as though Mother was a proud parent anxious to show off her family.

If our neighborhood had a flag, the pine marten would be a good symbol, rampant on a field of pine cones and wild raspberries. Here the marten population is in more balanced numbers than anywhere else because the old-growth coniferous forest is its indigenous and preferred habitat. In *The American Marten,* Denise Casey describes it:

> Martens do best in old forests. There towering trees and a thick canopy of branches keep the forest floor dark and damp, ideal for the many small animals martens eat. They also hide the marten from its predators and keep snow from piling up too deeply. . . . the number and kinds of trees and logs and shrubs and how they are arranged in the forest are just as important to the marten as how much food there is. . . . [It needs] hiding places, hunting sites, den sites and resting places. . . . The marten is an indicator species, a barometer of the overall health of the forest because they are so sensitive to change.

The pine marten is the only animal in this forest to be saved from extinction twice: once by the Russians half a world away and later by the growth of a new forest after its habitat had been destroyed by logging.

In the hierarchy of North American fur bearers, the beaver has always been king. During the great American fur century of the 1700s, there were many years when millions of beaver pelts were shipped to Europe annually, along with thirty to fifty thousand marten pelts. Fortunately for the beaver, the winds of fashion shifted and decreed that men's hats must be made of silk, and the beaver was saved. But the pine marten, the fur reserved for kings, noblemen, and high churchmen, was still trapped heavily, and the population was becoming dangerously low. Competing with the North American fur trade was the growing Russian fur industry, which featured such elegant peltries as the Arctic fox and the silky sea otter. They also had a marten (*Martes zimbellina*) in their line of furs, hardly detectable from the American pine marten except by a taxonomist. However, furriers from Siberia developed a method of fluffing or thickening the marten's fur, and marketed it as *sobol,* which became sable, the most luxurious and expensive of furs, and the bottom dropped out of the pine marten market in North America. *Sobol* made millions for the Russians, and it saved our pine marten from likely extinction—until 1921, when, after twenty-five years of logging and slash burning, the marten's habitat was destroyed here, and the animal was declared officially extinct. A half-century after its destruction, a new forest matured, and the pine marten was ready for another comeback. In the *Journal of Mammology* from February 1955, Milt Stendlund wrote: "An authentic record of a pine marten was noted on December 3, 1953, when Waino Starkman, of Ely, brought a marten pelt to local game wardens. The animal had been caught accidentally in a bobcat set near Burntside Lake. No authentic records of pine marten have been noted for at least 30 years in Minnesota."

How fortunate we are to have a stable marten population now. If I were to judge a beauty contest among the wild creatures, the pine marten would be my choice for the crown. I think it is the most beautiful animal in the forest: glossy chocolate coat, creamy-orange neck scarf, and a ballet dancer's grace and postures. It is the design perfection of the pine marten that distinguishes it from the other creatures. Its silky movements and aerial grace in the high pines are what make Wallace Stegner's description in *Where the Bluebird Sings to the Lemonade Springs* so true and vivid. The pine marten does "seem to be swimming along, even in casual motion, . . . a string of bubbles streaming along after it."

4
A Parliament of Owls

BIRDS OF MYSTERY and the deep silences, barred owls are our shadowy companions in this old forest. Likely because of its nocturnal habits, unblinking stare, and weird night noises that penetrate sleep and log walls, the barred owl remains a spectral presence, making its legendary role as death messenger or omen of tragedy seem quite believable.

At the Gunflint cabin we often hear its doleful requiem during the wilderness night, but seldom are aware of it during daylight hours. Last year a barred owl visited us regularly during the afterglow of the westering sun and gave us a rare glimpse of its menacing company.

Next to the dock an old white cedar clings tenaciously to the slabby granite boulders along the shore, its gnarled roots coiling and grasping holds in the fissured rocks. It slants out at a forty-five-degree angle to the water's surface, flat green sprays still growing at mid-trunk, but the gray, weathered, and branchless tip points out over the water directly at the North Star, like a warrior's lance. Each evening the barred owl glided in silently and perched at the end of the snag, watching the water below and casting an ominous shadow on the flotillas of young mallards and mergansers being shepherded to nighttime coves by nervous mothers. We often wondered if this sinister bird could have been Oley, an orphaned barred owl raised by our neighbor Peggy Heston, a tiny lady with a heart as big as a watermelon for the young of wild creatures.

Peggy had gone out to her mailbox one morning and found this frowzy little dirt-covered owlet squeaking in the gravelly row of scrapings the road grader leaves on the shoulder. At first she thought it might have fallen from a nest, as owlets frequently do because the adults are such slovenly nest builders. In fact, most of the time they just use the tattered remains of other birds' nests. Peggy thought if she left the owlet there, the mother owl might come and reclaim it, but when she went back in an hour it was still there, looking more bedraggled and forlorn than ever. She scooped it up in her apron and carried it back to her kitchen, where it perked up considerably after a substantial breakfast of chopped eggs, hamburger, and milk. She named him Oley, and he quickly adapted to the good life at the Hestons' house. He spent many hours watching television from a perch on the back of the davenport. When he got a little bigger, Peggy made him a sturdy roost, which was a bit higher so he would be safe from the dog, whom he delighted in terrorizing with sudden bizarre screeches.

Peggy kept Oley around the house for almost a year, although she knew that someday she would have to return him to the wild. She was concerned that the lifestyle he had become used to would spoil him and prevent him from working for a living, so in the spring she began his on-the-job training by dragging a stuffed mouse around the grass on a string. At first Oley thought this was a pretty dumb game and ignored the mouse, but his instincts were eventually aroused and gradually he began to get the idea, pouncing on the mouse and literally tearing the stuffing out of it. After several mouse replacements, Peggy decided Oley was ready for his finals. She took him out into the woods, released him from his tether, and went back to the house to worry about his chances of making it on his own after the pampered life he had been leading.

About eight o'clock that evening Peggy was startled by a commotion at the front screen door. When she went down to investigate, there, perched on the railing, was Oley, fanning his wings against the door and holding a deer mouse in his talons. We aren't supposed to believe that creatures can communicate with us, but it would be hard to convince Peggy that Oley wasn't telling her, "See, Ma, I can do it."

Perhaps its phantomlike reputation has developed because the owl is heard more often than it is seen. Its ability to remain motionless for hours and its elegant camouflage of gray, brown, buff, and black baffle even the most careful observer. And it has other tricks of concealment that often bewilder me even when I am certain there is a barred owl nearby. One of the most disconcerting is its ability to seem to become part of another object.

Every mature forest has an abundance of snags, the ghostly trunks of standing dead trees, broken off by wind or lightning, stripped of bark, and silvery with age. Many are spiraled with shelf fungus and dappled with the holes of insects and cavity-nesting birds. The barred owl finds the notchy tops of these snags perfect observation posts, where its mottled coat and blocky shape blend it into an upward extension of the trunk. There it sits, statue-still and unseen, until it suddenly glides off, pouncing on a scurrying vole or snatching a woodpecker off a nearby tree.

Hiking on the Loon Lake portage road one August morning, filling my bucket with the plump raspberries of late summer, I stopped to eat my lunch seated in a natural armchair formed by the upturned roots of

a windthrow, the exposed and contorted roots of a fallen white pine. Looking around absently in the moss nearby, I began to notice a number of owl pellets—the dried up and regurgitated remains of bones, fur, and other indigestibles that owls cough up and scatter near their roosts and hidey-holes. Because owls have no teeth, they cannot chew their food. They tear it into chunks or swallow it whole. The materials their body cannot digest are packed with a mucus in their stomachs and then upchucked. Seeing several of these pellets near the snag, I got up, stepped a few feet away, and searched closely. Sure enough, there, right at the top of the snag, was a barred owl, wings slightly drooped over the broken top of the snag; only the unblinking gaze of its molasses-colored eyes gave it away. The eyes are the quickest and most accurate field marking of the barred owl. All other owls here, except the great gray and the barred, have lemon-yellow eyes. They are about the size of human eyes and can change focus rapidly, but can concentrate light with about one hundred times the ability of human eyes, allowing them to zero in on their prey instantly. Unique, fluffy feathers along the leading edge of their wings enable them to fly silently and strike without warning.

The northern barred owl (*Strix varia*) has several other distinguishing features that make it easy to identify. It is a big bird, fourth largest among the eighteen North American owl species, and has a vertically barred breast; a round, tuftless head; and very large ear cavities, which give it the keenest hearing of any bird. In *The Owls of North America,* Allan Eckert reported a barred owl that could hear a mouse running on hard-packed earth fifty yards away.

But it is the voice, an exceptional voice, that gives it away even in the dark of night. The most familiar call is the four rapidly repeated "Hoo's," which sound remarkably like "who cooks for you?" But it has a variety of other sounds and phrasings, the most startling of which is heard during the late winter mating and is a bloody-murder sound that mimics a woman screaming. It can also hiss, chuckle, growl, and groan. One of my birding friends has her own voice identification formula. "If it sounds like nothing else on earth," she says, "it's a barred owl."

They fare well here in this forest. Voles, shrews, deer mice, perching birds, and an occasional snowshoe hare are on the daily menu, but consistent with its other eccentricities, it might wade a stream for frogs or even kill and eat a skunk because it has no sense of smell.

Another owl neighbor, far more delightful than the solemn barred owl, is a tiny clown, the saw-whet owl

(*Aegolius acadicus*). The common name sounds inappropriate now, but it came originally from the loggers who thought its monotonous, scraping voice sounded like the whetting, or sharpening, of a large mill saw. I prefer its baptismal species name, *acadicus*, given for the region of its first documentation, Nova Scotia. Its tedious, mechanical call is often repeated one hundred times a minute, and local birders have more appropriate descriptions of it, including a truck's backup signal or a spaceship in a swamp.

Just before starting the drive to the cabin one morning, I got a call from Orv Gilmore, who lives nearby, telling me to bring some batteries for my smoke alarm system. Its constant beeping, which signals weakening batteries, was keeping him awake. When I arrived, Orv was gone, but there was a note tacked on the door: "Sorry about the false alarm. The beeping turned out to be a lovesick saw-whet owl in the big cedar tree near the woodshed."

The owl's reputation for solemnity could not have been based on the behavior of the saw-whet. It is the most entertaining creature in the forest, and boreal show biz is its natural venue. One of its acts is a soft-shoe dance in which it hops up and down a branch, first on one foot, then the other, and it can turn its face upside down. Most owls can turn their heads 180 degrees so fast it looks as though they are turning their heads completely around. The saw-whet turns his head over instead of sideways, the only owl that does this. It is disconcerting to be focusing a camera or field glasses on one and suddenly discover its face is on upside down, the beak up near the top of its head and the eyes near its breast. It is also an accomplished ventriloquist, which makes it very difficult to locate. Even while staring at it, perched at the rim of its hole or in a thick bush near the ground, the voice will come from high up in the tree. It has virtually no tail, a seven-inch body, and a somewhat misshapen head, giving it a disproportionate, buffoonish appearance. Because the wing beats alternate, flight is erratic, setting it off on a staggering flight path. A thoroughly enchanting little bird.

Recently we learned that there is another owl here, but we have never seen it. It is the boreal, or Richardson's, owl (*Aegolius funereus*). The voyageurs gave it the liltingly beautiful name "la nyctale boreale," and the Montagnais Indians of the far north called it the "water-dripping bird." Both names refer to its ascending, liquid song, which is heard only on spring nights during the mating season. Research biologist Bill Lane tells me that that is why we have never seen it. "Boreals are strictly night hunters," he says. "And the boreal's call, a twelve-note rising trill that resembles a jack-snipe's territorial flight call, is only heard at night in March and April, which reduces the likelihood of birders spotting it."

The boreal owl was discovered here by naturalist Kim Eckert in 1978, the first sighting in the lower forty-eight states. Not long after, Steve Wilson, an ecological biologist, stumbled over one while trout fishing on the Baptism River. Since then, both Lane and Wilson have covered many miles conducting listening-post surveys and researching the elusive birds. Lane estimates that there are about 220 singing males in this area. It may be possible that I have seen one without recognizing it, because the boreal is very similar in appearance to the saw-whet. The boreal is a little bigger and has a sooty facial ring and a yellow beak, in contrast to the saw-whet's black beak.

There are other owls nearby in different habitats—cleared land, newer forest growth, and areas closer to Lake Superior where more moderate winter temperatures reduce the snow depth, making hunting easier. In these areas are the aggressive great horned owl; the great gray owl, largest of the North American owls, with its sixty-inch wingspan; and the American hawk owl, with its black sideburns and rounded, falconlike tail. Farther north in the tundra country are the regal

Arctic snowy owls, which sometimes migrate here in winter when the prey population crashes in their stark homeland, and the secretive long-eared owl, a dweller of the lonely spruce and tamarack lowlands.

"Night bird, death bringer, have you come for me?" asks Grandmother in Virginia Holmgren's fascinating tales of owl legends and folklore. Sir Edmund Spenser called the owl "death's dread messenger," and most American Indian tribes believed the owl was present at death and accompanied the departed on the last journey. During the Renaissance, artists often placed an owl on Calvary's cross in crucifixion paintings, and even the august science of taxonomy carries fragments of legend in its genus and species names for owls: *Strix* (Latin) and *Striga* (Greek) both mean witch or hag. The American hawk owl is *Surnia ulula,* bird of ill omen, and the boreal is *Funereus,* the funeral bird.

Why should all this mournful, dolorous symbolism be laid on the owl? Why the age-old association with death and tragedy? I suspect it is because most owls, except the silly saw-whet, are ideal metaphors for fear. They are mysterious and unknowable. They strike swiftly and silently with rapier talons. They make strange, haunting noises in the night that seem to come from shadow worlds, and their empty, inscrutable stare impales the imagination. Also, they have been around for years beyond counting, fifty million at least, in the same form. Every people in every culture have had owls among them, so the stories have had centuries to simmer in the juices of uneasy minds.

Of course, side by side with the frightening images are those of wisdom and sagacity. Those perceptions are why we call the group a parliament of owls, and we owe that mythology to Athena, the Greek goddess of wisdom, whose constant companion and adviser was a little round-headed owl.

The stories go on and on. Folklore is the vehicle and the result of the oral tradition's passage through generations, in the telling and retelling of indigenous stories. Along the way new creatures and new events work their way into the fabric of the stories, and so I think again about Oley and his stuffed mouse and the clownish little saw-whet. Maybe someday they will star in stories about how the barred owl learned to hunt and about the saw-whet owl who made the forest laugh.

5
The Magnificent Seven-Footer

ONE SUN-SPATTERED morning on the Clearwater Road, which doesn't have a straight or level stretch of more than a hundred feet, we were moving along slowly, listening to the drum-brush rhythms of pine boughs swishing across the hull of the canoe belted securely to the top of the car. In a deep valley to our right was a boggy pond of tannin-stained water ringed by heavy growths of sedges, cotton grass, and some stunted black spruces. In nodding clusters, back from the water's edge, grew pitcher plants, their waxy, burgundy blossoms bending low over the carnivorous pitcher-shaped leaves that trap and devour insects.

Ahead was a short, steep hill running up an esker that bordered the bog. As the road crossed over the top of the hill we came upon a startling spectacle. Next to an old pickup truck, a bearded and apparently demented young man in cutoffs was leaping up and down, waving his arms frantically and shouting, "A moose, a moose . . . I finally saw a moose!"

We looked down into the bog and there, seeming as puzzled as we were, was the object of the young man's hysteria: a young bull moose, his immature antlers dripping gobs of aquatic vegetation, knee deep in the coffee-colored water of the bog. When the man finally quieted down, he couldn't take his eyes off the beast. "Jeez, he's big. Just like I thought he'd be," he kept saying.

This little tableau was a great example of what we have come to call Moose Power, a mighty fascination with this whimsical and ungainly giant, the moose (*Alces americanus*). A full-grown male can stand seven feet high at the shoulder and weigh up to one thousand pounds.

Driving up the Gunflint Trail one day I noticed a car with Oklahoma license plates pulled off onto the shoulder of the road near one of the illustrated moose-crossing signs that look so official, but aren't. Remote country courtesy requires a stop to see if help is needed.

"No thanks, we're OK," the driver said. "We're just waiting for a moose to cross. We've never seen one, y'know."

These people were unaware that they were anticipating the unlikely. They were expecting the failure of an eternal truth: moose only appear unexpectedly. The hopeful watcher can park near a crossing for hours, prowl the forest roads near sundown, or hunker down in a blind and wait . . . but no moose will appear. However, go a little too fast on a curvy road, duck into the bushes for a moment, or complete a soundless stalk of a trout pool and there, magically, soundlessly, and instantly, will be a moose. And it will be staring myopically. "What are you doing in my space?" it always seems to be saying.

One May afternoon at the cabin, Julie was on her way out to fill the bird feeders when she came running back inside, calling to me excitedly that there were two moose standing next to the boat ramp. I thought she was kidding but went out to see anyway, and sure enough, there they were, a cow and her calf just out of the lake, shaking sheets of water off their slickened coats and staring at us insolently as though we were intruding. Julie, who always has more presence of mind than I, ducked into the cabin for her camera, came back out, walked up close to them, and snapped several pictures. Their modeling chores over, they gave a final watery shrug and ambled off into the woods.

No other animal around here seems to have the indifference to people the moose have. They don't bolt or disappear in a flash, like a deer or a wolf. They seldom challenge, except during the rut in the fall.

Sometimes they won't even get off the road, but just trot casually in front of the car, glancing back occasionally, until they feel like moving off into the woods. Some say they are not very bright—a moose's brain is only about the size of a baseball—but most hunters I know would dispute that. Another possibility is that a certain self-confidence comes with being the largest animal in the forest. The biggest kid on the block always goes wherever he pleases in the neighborhood.

That's your average moose; then there's the other kind. The kind canoe outfitter Bill Hansen from Sawbill Lake encountered on a forest road late one October night as he was returning from a meeting in town. Bill saw the moose standing in the road ahead of him, a tall, rangy bull with crooked antlers, and he slowed to a crawl, then stopped when the moose didn't move on. Eventually, after a little staring match, the moose trotted down the road and Bill started up again. But the moose's feint would have done credit to a wide receiver. Just when Bill got up a little speed, the moose whirled and charged. Bill stopped and ducked down on the passenger side to avoid injury while the moose reared up and came down with jackhammer blows on the hood and windshield. All was quiet for a moment. Then the moose gave the car a couple more clomps for good measure and stalked off.

That brings up Rule Number 1. When dealing with wild animals, they are always dangerous and unpredictable, even the normally docile moose, so always avoid any confrontation.

There are two separate moose ranges in Minnesota; this northeastern boreal forest and the farmland and patchy woods and bogs of the northwestern part of the state. There is an alley of about one hundred miles between them, and the moose never mingle. The farmyard moose, or Agassiz herd moose, are a little smaller than their forest-dwelling cousins, and they feed primarily on the bog willow, aspen, conifers, and grain crops they can snitch from the farmers. There have even been complaints from some farmers in that area of the moose grubbing up sugar beets that are grown there.

The moose, like the black bear and the timber wolf, is a longtime resident here. They all came to the post-glacial forest from Asia about ten thousand years ago across the ice bridge over the Bering Sea. Nature and the Indian hunters controlled the moose population for centuries, and it wasn't until 1865 and the signing of the Treaty of La Pointe, when white settlers came into this region, that the dynamics of the moose population were seriously upset. With settlers moving in to farm, log, and prospect for minerals in a region where no cattle could be raised, the moose became the principal source of red meat nutrition for many years, and they were slaughtered by the thousands. Hiram Drache reports on the situation in Saint Louis County in the 1890s in his book *Taming the Wilderness:*

> Moose were common in the virgin forest. The land had not been touched when Malker Finstad moved his family there and moose were readily available. They became the main source of meat. . . . Steve Mennausau shot 37 [near Loman]. . . . The steaks were stored in brine crocks, or ground and made into meatballs and canned with gravy. Sandwich meat was salted, smoked and dried on racks above the stove before it was stored.

Logging companies decimated the forest and the moose herd at about the same rate. They hired marksmen to feed the lumberjacks, and one professional hunter, who worked the Cross River region, killed thirty-five moose in one month.

Logging had another serious effect on the moose in this region when huge tracts of land were opened and the white-tailed deer moved in from the south and west, attracted by the new growth of aspen and birch. These deer brought with them *P. tenuis,* a parasitic nematode called the brainworm, which killed the moose but left the deer unaffected, one reason deer and moose shouldn't share the same range. By 1922 the moose population was so low that hunting them was prohibited, and the season was not reopened until 1971, and then only on a lottery basis, a system that is still in place today.

For all its size and strength, the moose has another minuscule enemy, the winter tick, a vicious bloodsucker that swarms over the moose in hordes, draining its blood and causing an agonizing death. Justine Kerfoot found a dead moose in the woods near here that she described as being "shiny" with the little assassins, almost as though it were varnished with them.

So the moose struggles on with its tiny enemies, trying to maintain its herd size in balance with the size of its range. It also has to stay wary of timber wolves, who take a certain number of moose each year, usually the sick, the elderly, and young calves. A full-pack attack is almost always fatal, but a lone wolf would seldom risk even being nicked by one of those flying hooves. Even though a blow may not kill, it could cripple the wolf, and in the perfect communism of the wolf pack, each member must contribute in order to eat. A crippled wolf can't contribute to the hunt, and the wolf pack has zero tolerance for noncontributors, who are killed or ejected from the pack.

To the Ojibwe the moose was Moozo, the twig eater, which describes its winter feeding strategy—browsing the tips of conifers, aspens, and willows. It has to make the most of every feeding possibility because it needs about sixty pounds of food each day. In the spring, hardwood bark is a choice morsel because of the rising sap, along with the new aspen and willow twigs, with their nutritious bark and catkins. The summer woods provide ferns, mushrooms, and horsetail, but summer finds the moose spending much of its time in the water. One of the reasons is that the water offers relief from the tartarish hordes of mosquitoes, black flies, and other insects, but there are also lots of moose goodies there—lily pad roots, bladderwort, and pondweed, all sodium-rich and nutritious. All this eating, combined with the rich milk of mother moose, produces the fastest-growing animal in America; the calf gains approximately two hundred pounds in its first three months. In *The World of Moose,* Joe Van Wormer reports that the moose has an uncanny ability to know exactly where in the water the plants it wants are located, even though the water may be very deep and nothing is visible on the surface. He has seen them dive as deep as eighteen feet to get the right aquatic vegetation.

The moose provided the Ojibwe with a veritable general store, including a pharmacy. The hide was used for moccasins and snowshoe webbing, the meat for sustenance, back sinews for sewing thread, antlers and bones for tools, the bristly mane hairs for embroidery, and the hooves for rattles used in religious ceremonies. The pharmacy department was heavily used. The feet were ground and boiled in a decoction for epilepsy, the antlers were sliced and boiled to produce a snake-bite remedy, dried nerves were used to treat arthritis, and the fat was rendered and used as a salve for sore muscles. Some Indians would kill a moose and move their camp to the kill site, where they used up the whole animal, and would then move on to the next kill.

To me the moose's most outstanding and endearing characteristic is its built-by-committee appearance. It is this strangeness that allows A. L. Karras, in *North to the Cree,* to justifiably use words like "grotesque," "awkward," "fantastic," and "magnificent" all in the same sentence when describing the moose. It has a monster nose, called a muffle; stilt legs with knobby knees and great clodhopper feet; a dewlap, or bell, of skin and hair hanging from beneath the chin of both sexes and of absolutely no use; a tiny little tail; and huge eyes; and, during the mating season, the bull has green hair growing between its toes. This staining is caused by a pheromone that is secreted during the rut. The moose has a weird foghorn-like voice that is heard in the autumn swamps when the moose goes a'courting.

So add whimsy to all these descriptions of the moose, . . . and yet, coming upon a full-grown bull standing at the water's edge while canoeing through the early morning gauze of a misty bay is a heart-stopping experience. The great head rises slowly, antlers glistening. As he tenses, huge muscles ripple and sweep along under the almost-black coat, and he stretches up to his full magnificence. That is a moment that imprints forever on memory.

6
Rascal in a Gray Suit

THE GRAY, OR CANADA, jay (*Perisoreus canadensis*) and I have some things in common: we prefer the boreal woodlands, we both love to eat, and neither of us can sing very well. The gray jay has a variety of calls, most of them strident and unpleasant, but it can purr like a cat when it is begging.

Some bird writers are impressed by their vocalizing. Charles Bendire wrote, in *Life Histories of North American Birds,* "While some of their notes are not so melodious, I consider this species a very fine songster." I have never heard a gray jay I considered a fine songster. Mostly I hear complaints, screeches, scoldings, and shrill screams much like the red-shouldered hawk's.

A somber-looking bird, smoky gray, with a monk's black cowl and a white breast, the gray jay is a little larger than a robin, with the long powerful wings of a soaring bird, but I have never seen one in long flight. When they approach a feeder, they glide in a series of short flights from higher to lower branches, then reverse the process when leaving, flying ever higher in the branches. Frank Shunn says gray jays only fly high during early morning hours. He claims they are up there looking for chimney smoke from the cabins, which means someone is up making breakfast. The jays home in on the smoke signal and hang around until someone comes out with leftovers.

Other jays, such as the blue and Stellar's, can usually be found on cultivated land and the lawns in town, but the gray jay thrives best far from settlement in northern latitudes. Its insatiable appetite marks it a Corvid, along with other family members—crows, ravens, and magpies, the gluttons of the bird world. Author and ornithologist Laura Erickson offers some insights on the gray jay's appetite in *For the Birds:*

> Gray jay's diet gives a new meaning to the word omnivorous—they've been reported eating moccasins, fur caps, matches, soap and plug tobacco. They assiduously followed the sleds of early trappers, sometimes lighting right beside the trapper poking holes in his pelts. Gray jays carry off anything remotely edible, including candles.

Its nicknames, "camp robber" and "meat hawk," testify to its endless and eclectic appetite. First it begs and wheedles, and if that doesn't work, it steals. Another nickname, "whiskey john," is a corruption of its Indian name, Wiss-ka-jon, and eventually became "whiskey jack." In spite of its garrulous behavior and gross appetite, the name doesn't come from its drinking habits.

For fascinating reading, the old bird books far outclass the new data-filled field guides. One such gem is *Birds of Canada,* by P. A. Taverner, who says of the gray jay (he calls it the Canada jay, of course):

> If the other jays are clownish, one scarcely knows how to characterize the Canada jay. It has all the family characteristics in an exaggerated form, but seems to lack the keen appreciation of its own humor that the others possess. Its entire lack of self-consciousness, or poses, is notable and it does the most impudent things with an air of the most matter-of-fact innocence. . . . even the tent flap has to be kept tightly closed against its sharp eyes and inquisitive bill.

Unlike most passerine, or perching, birds, the gray jay is about 80 percent carniverous. Insects, worms, grubs, caterpillars, the young of other species, and mice are particular favorites.

Their nests are substantial conical structures of small sticks and plant fibers, lined with animal fur, feathers, or moss and almost always about ten feet up in a conifer. The nest needs to be warm and protected from the high winds of late winter because the three or four eggs are laid in mid-March and hatch in early April. The sooty black young grow very quickly so they will be well developed and ready to kill and eat the nestlings of songbirds that hatch early in June. Not a very endearing trait, but Nature devises some brutal strategies for the survival of her children. In the winter, when very little protein supplement is available, they follow the large predators. Wolves and man, being the two most effective predators here, get most of the gray jay's attention in winter.

In this northern forest winter days sometimes get very long, with their thin daylight hours and heavy snows that keep cabin dwellers close to the fire, but the gray jays can usually be counted on to liven things up even in the most inclement weather. They know the location of all the feeders and what is likely to be found on them. Everything is on their wish list: seeds, corn, suet, stale bread, peanut butter, scraps, anything they can get away with. Confronted with this feeder bonanza, the gray jay makes many trips to a variety of storage spots; the crotches of small trees and the rows of woodpecker holes in dead trees seem to be the favorite food caches. They return comedy and companionship for their larceny, an eminently fair barter.

One of our first winters here was spent during a time of expanding recreational facilities in the county, and a feature in our neighborhood was the brushing out, grading, and spiffing up of the East End Trail, which runs through a green tunnel of spruce trees on the Canadian side of the lake along the abandoned grade of the old Pee Dee Railroad, which was built to haul iron ore from the Paulson Mine at the end of Gunflint Lake to Thunder Bay, Ontario. The railroad was never used and the whole cockamamie scheme collapsed, but a beautiful cross-country ski trail is now the unexpected memorial to the project.

On a crystalline January day, Julie and I skied across the lake toward the Magnetic Channel, into Charlie's Bay, and through a cedar swamp and then headed east along the new trail. At the one-mile post, Jim Thompson had built a slab-sided, open-ended shelter in a grove of balsam fir trees on a little point overlooking a pond, a perfect wayside rest and lunch spot. We hadn't even gotten our ski poles stuck in the snowbank before three gray jays arrived (for some reason, they always seem to travel in threes) and began inquiring about the lunch menu. They were splendid-looking birds, elegant in their gray, black, and white finery—and for the moment, on their best behavior. No sooner had lunch come out of the backpacks than one hopped right up on my shoulder, peeking at the unwrapped sandwich and watching closely as I took my first bite. Crumbs that fell on the snow were immediately snatched up by the other two, but my new friend stayed put on my shoulder, hoping for a bigger bite. Of course, I shared my lunch with it. I couldn't resist its imploring looks and the chutzpah of its approach.

Their friendship seems to be of the fair-weather variety, because they disappear immediately after the food is gone. This day on the East End Trail they did follow us for about a half mile just in case we might change our minds. Then they left, and we didn't see them again for several hours, when we stopped for a rest near the end of the trail before starting the long ski across the lake and back to the cabin. Again, no sooner had we taken off our skis and sat down on a log than they reappeared, raucous, up close and personal, ready to eat again.

It seems that this gray jay behavior has always been consistent. In 1946, Arthur Cleveland Bent recalled similar experiences with this jay in *Life Histories of Jays, Crows, and Titmice:*

> Curiosity is another characteristic of this jay. One can hardly enter the woods where these birds are living without seeing one or more of them; the slightest noise arouses their curiosity, and they fly up to scrutinize the stranger at close range, often within a few

> feet, and they will then follow him to see what he will do. The sound of an axe always attracts them, for it suggests making camp, which means food for them; and the smoke of a campfire is sure to bring them in.

Our old-growth neighborhood around here is perfect gray jay habitat, but several miles to the south and west of the Gunflint Trail, where logging skinned the hills bare, the second-growth forest is about one hundred years old and is a mixture of sugar maples, balsam, yellow birch, paper birch, black ash, spruce, mountain maple, and aspen—still good gray jay habitat, but getting close to the southern part of their range, as they occur only irregularly south of Duluth.

In this deciduous hill country closer to the Big Lake, hidden behind a long sloping ridge, there is a small cabin and a beautiful large garden in the middle of a fifty-acre woods. It is a very private place, the kind Finnish people call *louastari,* or sanctuary. The view down the south-facing slope toward the garden reveals a kaleidoscope of blossomed color blocks. Brilliant reds, purples, and yellows explode over lower-growing beds of milky bellwort and Iceland poppies. One hundred twenty different perennials and twelve annuals scatter their dazzle over and around the vegetable plots and berry bushes. The fruits, blossoms, and seeds from this Eden in the wild create a bonanza for migrating birds, bees, and butterflies.

In recent years the gardens have become a health spa and rehab center for ailing and infirm birds of all species. Recently Laura Erickson, a licensed bird rehabilitator in Duluth, got a call from a conservation officer who had confiscated two injured gray jays from an illegal trapper who captured wild creatures for pets. Remembering the gardens and the owners' commitment to wild things, Laura called to see if they could take two gray jay patients. Of course they could.

The two young jays, who quickly became Jack and Jake, were badly debilitated. Feathers were worn off and muscles were atrophying from abusive confinement and poor care. Their caretakers, Jean and Gary, made room for them in the garden toolshed, and the jays quickly made themselves at home, poking into everything, hiding food in the rafters, pooping all over the workbench, and devouring a mouse that the cat had caught and reluctantly contributed to the patient tray. Jean and Gary were afraid both jays were quite far gone, but after a week of the good life they appeared to be at the top of their game. Releasing wild birds that have been sick or injured is a chancy thing. If they aren't in good shape, they can't compete for food and are easy prey for hawks, owls, or other predators. After a close examination, Jean and Gary decided to set them free. They removed a window from the shed, and Jake and Jack, with a new lease on life, darted through the window, crossed the garden, and streaked into the woods without so much as a backward look or a dip of the wings in salute to their benefactors.

How like gray jays! Eat everything in sight, complain, and then leave. So far they haven't been back, and no one expects them. I'm not so sure. Gray jays never forget a free meal.

7
The Empire Builder

THERE IS A BEAVER pond nearby that is bisected by a gravel fill road we check out often in the early summer, looking for feeding moose. It is a small pond, maybe six or seven acres, with a beaver dam and lodge on one side of the road and open water on the other; it's surrounded by flood-killed spruce, splotched with dead stumps, and fringed with cattails.

There is something growing on the bottom of this pond that moose like, probably pondweed or bladderwort, because when we go that way we often see moose wading on the open-water side of the road. Usually there will be a cow with her spring calf belly deep in the quiet, iron-stained water.

One day as I slowed my car to a crawl, approaching the pond cautiously, I noticed a beaver (*Castor canadensis*) sitting on its broad, flat tail at the shoulder of the road. Two moose were in the pond pulling huge gobs of vegetation from the bottom. I parked off the road and walked as slowly and softly as I could toward the beaver. The beaver has very poor eyesight for a wild creature, but makes up for that deficiency with extremely keen hearing. I got within about fifty feet of it and it hadn't slid back into the water, so I sat down to watch the moose with it. We were like two strangers watching a tennis match, not speaking or acknowledging one another but each casting an occasional sideways glance at the other. Not friendly, but not hostile, either.

I soon became more interested in the beaver than I was in the moose. It was grooming itself in the way unique to beavers. Each foot has two toes, or combing claws, on the inside, which act as clamps to squeeze oil from a set of glands on the abdomen. The beaver then spreads the oil over the shiny fur and combs it through the hair with these double claws. In this way it keeps its fur unmatted and water-repellent, as well as creating an effective insulation.

This beaver seemed so innocuous and mild-mannered, even a little prissy. Peter Matthiessen called the beaver a "sedentary, civic minded rodent." In *The Beaver Men,* Mari Sandoz claims, "The beaver is perhaps the most orderly and responsible creature." And yet this roundish, modest little neighbor sitting and watching the moose with me was Beaver, the Empire Builder, as Minnesota historian Solon Buck has called it.

It is mind-boggling to think that this diffident creature here at the swampy roadside was responsible for an entire segment of the western world's economic history for over a century. The beaver trade created the greatest economic boom the country had yet seen and the largest industry on the continent from 1670 to 1825. H. Silver, in her *History of New Hampshire,* says, "The beaver stands high as a figure of importance in history and perhaps should share with George Washington the title, Father of his Country."

Much of this beaver trade came through the Boundary Waters in the great canoe brigades of the voyageurs, right past our cabin on "Flinty Lake," as trader John McDonnell called it in his 1793 journal. It is estimated in *Furbearer Harvest in North America* that approximately one hundred thousand beaver pelts came through this region each year for many decades. The beaver trade's preeminence in the early years of Minnesota's economic history is indicated by the fact that nine counties in Minnesota have been named for fur traders: Aitkin, Faribault, Morrison, Olmsted, Renville, Rice, Sibley, McLeod, and Brown.

An even more incredible sidebar to the beaver story is that all this nation-building was focused on a fashion item, a whim of vanity—the gentleman's beaver felt hat. The method of combing the soft underfur from the beaver pelt to press into felt had been known for years, and by the 1600s, "kings, princes, nobles, wealthy merchants and even senior army and navy officers wouldn't dream of appearing in public without wearing a beaver hat," we read in David Thompson's diary, quoted in James K. Smith's biography. Beaver hats became so required by fashion that in 1638, King Charles I of England issued a decree stating that all men's hats must be made of beaver felt.

By midcentury, when the mania for beaver hats was at its peak, the coureurs de bois, or bush lopers, had begun to push north and west in search of the furry gold. In *The Beaver Men,* Sandoz says, "All across the continent beavers slept in the sun on their houses or cut the glassy surface of the pond with their whiskered noses as they swam and played and worked, but to the east a great fur trade area had opened and the race of three nations to build empires on beaver hides had begun."

By 1700 the beaver was extinct in all of Europe (except for small, remote sectors of the Scandinavian peninsula and Siberia), in eastern Canada, and in New England, and trapping was moving west through the Great Lakes region, northwest into the Athabasca country in northern Saskatchewan, west through the Rocky Mountains, and south as far as Arizona. All these millions of beavers were either shot, clubbed, or caught in crude handmade sets; the steel beaver trap wasn't invented until 1821. But as fashion had commanded its extirpation, so eventually did it save the beaver from total extinction. The wind finally shifted in that volatile world of fashion in 1824, the market collapsed, and the beaver was saved.

Now this neatly groomed little beaver sitting near me in the fiery orange hawkweed at the roadside is relatively safe to raise a family of two or three kits a year in that well-engineered lodge of peeled logs and sticks it has built near the dam. There is still a beaver-trapping season each year, but now there are controlled seasons and limits, and beaver trapping is more of a habit or a hobby. The trappers now go out in snowmobiles and pickup trucks, and their furs will barely pay for the gas they burn. Some of the pelts are sold to furriers, but others are tanned and sold to resorts and gift shops for tourists, who take them home as mementos of their visit to the land that provided parts to the economic engine that built this country.

The imprint of three and a half centuries of beaver trapping has produced a whole folklore, with legendary figures of almost Paul Bunyan stature, such as Crazy Hjalmar and the Three Mad Trappers from

Rat River. They were swamp ghosts—woods-smart, tireless, and independent. Often they lived on stale bread, jerked venison, and lake water for days in terrible weather. They moved through the woods silently, unable to build a fire or leave any trail for fear of attracting the attention of the game wardens who were almost always right behind them. These men were often poachers, outlaws, and their greatest joy, other than an unmarked blanket beaver pelt, was matching wits with the game wardens. The two weren't really enemies, just opponents in a high-stakes game. Often at the bar they were friends and had great respect for each other. The poachers' highest honors went to Charlie Ott, veteran Grand Marais game warden. He was a relentless pursuer and would go into the woods afoot or on snowshoes, never coming out until he had his man. "He was a tough old sonofabitch," they said of Charlie, the trapper's ultimate accolade for a game warden. There was an accepted code of behavior that none ever violated. One time a poacher with a grudge repeatedly dumped gut piles on the front doorstep of Warden Bill Hansen's cabin near Winton. The other trappers got together, found out who the sorehead was, and took care of the problem their own way. No questions were asked.

In those days, Winton wasn't much more than a boathouse on the Shagwa River, a few cabins, and twelve saloons a few miles from Ely. The lumbering days were over and the big payrolls gone. Underground mining had started at the Zenith and Pioneer mines in Ely. There was some commercial fishing on Basswood Lake and a few tourists were starting to come, but times were still tough, and a Duluth pack full of prime beaver hides plugged a big hole in the family budget. Getting them, storing them, and selling them, all under the gaze of the game wardens, took a lot of creativity and what the Finns call *sisu;* "guts" would be a close translation.

Perhaps the most colorful of the trapper/poachers (the words are used synonymously) was Leo Chosa, whose ambition was to go to Oklahoma, marry a rich Indian woman, and retire to the Forest Hotel in Ely to write a best-selling novel. This is how we are introduced to Leo in Peter Nowak's oral history of the area:

> In spite of Leo's reputation for high living and beaver trapping, he was known as a very devout Catholic. People commented on how often Leo was seen going to Mass. The game wardens knew that Leo had a big cache of furs somewhere, but no matter how hard they searched, they never found a thing. Much later it

was discovered the reason for his frequent trips to church was because the furs were stashed in a small enclosure under the church stairs.

For his project, Peter Nowak taped conversations with old-time guides, trappers, and loggers. Among these treasures of old north country tales are interviews with people like Leonard, an old trapper who grubbed a living out of this harsh, bush country for many years during the twenties and thirties. Leonard tells Nowak of one adventure with his neighbor Stone and how he got caught with illegal beaver hides in the Parent Lake country north of Tofte.

> Instead of changing places, he [Stone] used the same camp over and over again. . . . that was his mistake, using the same camp. He had six beaver skins, blankets we called 'em, in the packsack. That's all we ever took. Anyhow, he made a fire for coffee, and he left tracks too. That's when they come up on him . . . Ott and Dalbeck [game wardens from Grand Marais].
>
> I was coming through just to bid him the time of day when I seen what happened. I didn't go no closer . . . just watched. . . . It's 27 miles out from Parent Lake to a road, so I knew they was going to have to camp out overnight . . . but I followed 'em. They had the goods in the packsack and I've got just one thought in my head . . . if I can get that packsack, there isn't a damn thing they can do to him. So

I followed and watched. They finally made supper and Stone is eating with them, but they're watching that packsack pretty close. It was about 9 o'clock and they were getting ready to sack out. They had sleeping bags but Stone only had a chunk of canvas like me to roll up in. Jesus, we lived like dogs in those days.

Well, I waited about another hour and then I crawled on my belly for about three quarters of a block and got hold of that packsack. When I got it, I just took off through the woods . . . made a helluva noise, but being dark, they couldn't follow me. They waved their flashlights around, but I was long gone . . . kept right on traveling and hid that packsack near an old logging road.

Jesus, they had to turn Stone loose because they had no evidence and they were mad. Later I went to Stone's cabin and he's feeling pretty bad, but he didn't want to say nothing. He's sitting there and I'm talking about the weather and hunting and finally ask him who he's going to sell his beaver to.

"By Jesus, Leonard, that Ott and Dalbeck got me. I got nothing to sell. I got a whole year with nothing to eat but venison and fish. It's going to be hard for me."

I asked him how come they didn't take him in . . . and then I started to laugh so hard I almost had a fit. He squinted hard at me. "You son of a bitch, was that you?"

"Yeah, I got 'em back in the woods," I said. He was so happy, we went down and bought four quarts of whiskey and two cases of beer. We had 'er made.

A creature as storied as the beaver is bound to have many myths surrounding it, and many of them have stuck. One is that the beaver carefully notches the trees it is felling on the downslope side first so they will fall toward the water. Not so. The beaver is a smart builder, but it hasn't really gotten the knack of felling trees efficiently yet. Beaver-cut trees fall in all directions, sometimes killing the beaver as they come crashing down.

The beaver's tail, which is sometimes used as a sit-upon, is a wondrous appendage that performs many tasks. It is attached to the beaver by a swiveling bone for power swimming so it can be used as a sculling oar when the beaver is pushing a big load. It is not used as a trowel to carry mud to the construction site, as some early beaver books claimed, but it is a handy kickstand. The beaver props itself up with it while gnawing on a tree or when eating. At the instant before it dives, the beaver whaps the water with its tail, making a loud noise and sending spray flying like spindrift. There may be other reasons for this, but it is most likely to warn other beavers that there is danger nearby.

Beaver meat is said to be a delicacy, but the tail was especially prized by Indians and trappers. Early Jesuit missionaries had a neat rationale for eating beaver meat on Friday or during Lent. They claimed that the

beaver lived in water and ate fish, so therefore was a fish by identification, if not in fact.

The beaver has two large castor glands, scent glands, inside the lower gut, which secrete a gooey substance called castoreum, a marvelous matter that has been widely used in folk medicine since the time of ancient Greece to the present, and also as a base for perfume. The castor glands sold as well as the pelts during the fur century, and the Hudson's Bay Company used to ship an average of twenty-five thousand pounds of them to Europe each year.

The Anishinabe believed that in ancient times beaver were people and could speak. The legend says that the Great Spirit, who also created man, eventually took the power of speech away "because the beaver was becoming superior in understanding to man."

A myth, of course, but one day at her secluded cabin, poet Joanne Hart wasn't so sure. Joannne lives in a very remote section of the county at the end of a dirt road on the bank of a turbulent river. The surrounding country is heavily forested, ridged and scarred by torturing glaciers, still grimacing in a series of sharp fissures, canyons, headlands, and projections through which the mustang river plunges wildly, dropping over a thousand feet in twenty miles down into Lake Superior. There are a few low, flattish places where the river settles down and seems to rest and gather itself before the next lurch lakeward. One day, after skirting the river on an old Indian trail through a fragrant grove of spruce and wild fruit trees, Joanne found an ideal bathing pool where the river has cut a placid little inlet over sand and gravel.

Floating around in the pool one morning, enjoying the summer sun coming over the notchy cliffs on the Canadian side of the river, she was suddenly aware of voices in conversation nearby. She swam, waded, and scrambled as fast as she could for the flat rock near shore where she had piled her towel and clothes. Crouched behind alder bushes, she dressed and listened carefully, trying to locate the voices. They seemed to come from a spot where a little creek entered the river a few hundred feet upstream. The voices didn't seem to advance or fade, but the conversation was quite spirited. Thinking perhaps they were picnickers or fishermen at the creek mouth, she began to work her way quietly up the rocky shore. When she got close to the creek, she hid in the brushy branches of a blowdown red pine and peeked through the needle clumps at the widening of the creek where it entered the river. No picnickers. No hikers or fishermen. But there were three beavers in midstream and one sitting on shore carrying on a lively conversation. Either they finished their discussion or else sensed Joanne's presence, because the big beaver on the bank pitched into the water and the others followed with tail-slapping surface dives.

Joanne was dumbstruck. She couldn't believe she had heard this quite animated conversation among four beavers. Maybe this ability to communicate is not yet well enough studied, or perhaps the Great Spirit overlooked these four, hidden away, as Joanne is, in this remote northern wilderness.

8
Along Came a Spider

PERHAPS DON MARQUIS was right when he said, "Beauty has the best of it in this world. . . . the only insect that succeeds in getting mourned is the butterfly, while every man's heel is raised against the spider."

I can't seem to find the word in my dictionary, but arachnophobia—a fear of spiders—seems a logical construct. The arachnids are a large family, including mites, ticks, and chiggers, but none of them inspire the fear and horror that spiders do. Especially *Lycosa,* the wolf spider, who sends the arachnophobes shrieking and running for their baseball bats. What is that revulsion all about? Ugliness, I think. *Lycosa* is definitely ugly to many people. Huge, blotchy, and hairy-legged, it stares out at its world from eight malevolent eyes. With its legs fully extended, it can easily straddle the rim of my coffee cup.

I think they are really quite handsome creatures, at least the ones that wander around our cabin hunting mosquitoes and flies along the windowsills, under the bark of firewood, and in the dark recesses of bookcases. They make tractable pets, are very tidy, and never bite—but try to convince the arachnophobe of this.

Ron Hemstad is not arachnophobic, but he does have some hard feelings toward *Lycosa*. His is a longstanding grudge that started many years ago at the Hemstad cabin in the middle of a quiet summer night. His daughter Nancy was sleeping up in the loft when she awoke suddenly with a strange tickling sensation on her tummy and discovered a large wolf spider strolling along under her nightie. The clamor and screeching that followed this discovery has probably biased Ron's attitude toward this 300-million-year-old gentle giant of the spider family, which has found our north woods environment much to its liking. Of the two thousand wolf spiders that have been classified, only about one hundred live north of the Mexican border and only a few of them live this far north in the boreal forest.

One summer, after the nightie affair, Ron was installing some new flashing around the chimney on their sharply pitched cabin roof. A dangerous place to work, but he was tethered by a rope around his middle that was anchored over the other side of the roof. Inching slowly upward, he was getting near his goal when he glanced up at the point where the ridge beam met the chimney and stopped short. There, glaring down at him, just a few feet from his face and standing her ground, was the mother of all wolf spiders, pedipalps weaving hypnotically and jaws snapping. Thinking this was probably a relative of the one who had been cavorting in Nancy's nightie, Ron had a few anxious moments.

Upon being summoned loudly, his wife Betty was overjoyed at the spider's presence, which didn't improve Ron's mood at all. Betty is a wildlife photographer with exquisite sensibilities, and to her the monster, the most photogenic of all spiders, presented an opportunity only an artist could long for. As stable and rational a person as I know, Betty is completely convinced we can communicate with wild creatures through concentration and the proper attitude. She ducked back into the cabin for her camera, leaving Ron and the spider playing chicken on the roof. Returning with her equipment, she climbed the ladder, scrambled up the roof with a cardboard box, and set it as close as she could get to the spider. After several moments of silently conveying her wishes to the spider, it graciously marched down the roof and climbed into the box, leaving Ron alone with his thoughts.

Returning to the ground with her prize, Betty considered an ideal setting for her picture, finally settling on a nook in the cobbles near the stream where she placed some maple branches and the spider. For several minutes she busied herself with depth-of-field considerations, light measurements, and all the precise fussiness of the careful photographer. When she finished shooting a roll of film, she communicated to the spider that the sitting was over, and it stalked off into the woods. Ron has a slightly different story of the episode, but Betty has a startlingly beautiful set of pictures.

Lycosa is a fearsome-looking but fascinating creature, although not all my family share that view. One day when an adult male was prospecting in our kitchen drain strainer, Julie accidentally turned on the hot water full force, and the poor thing curled up like a moth hitting a flame. When I pointed out that that was a pretty insensitive thing to do to a wild creature, she didn't seem very contrite. Her attitude was that the wolf spider was not exactly an endangered species around our cabin, and if I wanted to study one more closely, I could take a good look at the spider on the lamp shade in our bedroom. Peering at that one closely revealed that she was a mother, with eight little silk papooses woven of white silk plastered firmly to her carapace and abdomen. This habit of toting the silk-wrapped eggs is unique to the wolf spider; the mother carries the eggs in globe-shaped cocoons until the spiderlings hatch. If they accidentally come unglued or fall off, she scurries around until she finds them and pastes them back on. The sack is made of layers of silk that she spins out and mixes with mud. After the spiderlings hatch out they ride around on the mother's back for several days. For its usual winter quarters, the wolf spider digs a tunnel about six or eight inches long, slanting downward in soft ground, and lines it with spun silk for insulation. The spider then weaves a trapdoor, which it seals shut, and hibernates for the winter. The extended family of wolf spiders in our cabin have life much easier. They crawl behind the ceiling boards into the woolly insulation bats or under the wood box near the stove, and those who make it through the winter rejoin us in the spring.

Like the owl, the spider has played many roles in the world's folklore, and for centuries it was considered a beneficial creature. Its good-guy reputation likely started when it was credited with saving the life of Muhammad by spinning a cover over the mouth of a cave and concealing the prophet from his enemies. After that it was always connected with good fortune. In England it was believed that a spider seen crossing

the path of a bride and groom assured prosperity for the marriage. In Egypt this was not left to chance, as well-wishers placed a spider in the marriage bed. In Portugal the spider was called the "money spinner," and in Jamaica it was believed that seeing a spider in the morning meant that a visitor bearing gifts was coming.

Among American Indians, the spider was most often associated with magic. It was said that warriors killed in battle ascended to their new kingdom by means of a golden rope of wolf spider silk connecting earth and sky.

It was in the field of folk medicine, however, that the spider and its products made their most significant contributions, being used in various applications for the treatment of leprosy, hemorrhoids, earaches, smallpox, and the plague. It was a common practice in early British surgery to place wadded-up cobwebs on warts and set them afire. This was a successful wart treatment, probably because no patient ever returned for a second application. According to an old *Connecticut Journal of Natural History,* "It would seem that a creature which has enjoyed such a fine reputation as a benefactor of mankind would be better received than it is today, but the wolf spider doesn't seem to bask in the world's grateful glow."

In addition to the heel of man, its tormentor, the wolf spider has many enemies: climate and its cannibalistic neighbors kill most of them, and shrews, wasps, frogs, toads, and birds are constant threats. But *Lycosa* stands tall at its doorway, defying the world and ready to take on all detractors. It pleases me to see the beauty of wildness in this maligned but remarkable creature.

9
Helen Hoover's Deer

MANY GENERATIONS of white-tailed deer (*Odocoileus virginianus*) have passed this way since author Helen Hoover lived in the cabin next door, but her spirit must watch over them because they are still here, even though some biologists have been wary of their future for years. The old-growth forest just isn't deer country, they say. This is moose country.

There haven't always been deer in this region, and even after they came, their populations had their ups and downs. It wasn't until the end of the logging era, from about 1910 to 1920, that the first deer population explosions occurred. The great clear-cut tracts began to refoliate with deer browse species such as aspen, birch, and mountain maple, drawing great numbers of deer from the south and west. Now as the forest approaches maturity once again, reaching the conifer end of its succession, deer numbers are declining. This is good news for the moose, which prefer the old coniferous forest and don't usually share range with deer. There are still enough clear-cuts to provide deer forage, and many residents maintain corn feeding stations, so there will likely always be some whitetails here.

The game biologists don't think our corn piles are a very good idea. We're only making pets of the deer, they say. We are well motivated, perhaps, but it is not good ecology. It certainly doesn't do the moose any good to have the deer around. They are intermediate hosts to the nasty little roundworm called *P. tenuis,* the brainworm that destroys the moose's nervous system and eventually kills it. I understand the biologist's point of view, but I thoroughly enjoy watching the deer nose around near the cabin, waiting for me to scatter some corn on the snow. On any winter morning we can see a dozen or more stepping warily along the game trail out in back on their way through the old Hoover property.

Ade and Helen Hoover moved here in 1954 when they were overpowered by a dream many have had since. They gave up their jobs in bustling, crowded Chicago and moved to a cabin in the wilderness. They were picnicking on the lakeshore while on vacation one summer when the inspiration came to them. "Let's not go back," they said. And they never did. Helen, a metallurgical chemist, and Ade, a publishing company art director, gave up their lucrative jobs for the solitude and beauty of a primitive cabin on Gunflint Lake. Helen wrote children's books, magazine articles, and adult romances (under the pseudonym Jennifer Price). Later she produced the nature classics *A Place in the Woods, Gift of the Deer,* and *The Long Shadowed Forest.* Her last book about the area, *Years of the Forest,* was written in Taos, New Mexico, where they moved in 1973. Ade did all the pen-and-ink illustrations for her work, as well as seeing to the endless repair on the old cabin, which had been built by trapper Walt Yocum, a Kentucky woodsman, who had moved here with his wife, Addie, in 1936. Walt built it of spruce logs he had found cut and piled in the forest nearby, using the twin gable or cricket design favored by generations of Kentucky cabin builders. It turned out later the logs had belonged to a neighbor, but the cabin had been built, so the neighbor graciously gave them to the Yocums as a housewarming present.

Helen was a relentless preservationist who loved the wilderness setting and the scarcity of close neighbors. She and Ade cherished the animals and encouraged all of them to make their homes in or near the cabin. They fed the mice rather than trapping them, and the red squirrels would run across the floor and up the walls unafraid. She named most of the deer that visited regularly, gave them oatmeal in the morning, and worried herself sick if they didn't show up. She kept her darkest thoughts for hunters, whose red caps punctuated every acre of the neighborhood during deer season, and even threatened a few of them with Ade's old Navy .45 when they cut through the Hoover property.

In the winter they planted a garden of cedar boughs in the deep snow when the deer could no longer find browse and fed graham crackers to the red squirrels in the kitchen. Helen was not a gregarious person, according to those who knew her well, but she was a woman of great compassion for wild creatures, and she wrote of them with intense feeling and a surprising depth of knowledge for a self-taught naturalist. Her books sold in the millions, and autograph seekers who had no respect for the Hoovers' privacy bothered them constantly. But with financial independence in 1967, Helen began to write about a felt "need to get out and look around for a place to continue our life and our work." Later she often spoke with great longing about the "timeless healing of wilderness," and her feeling of being "whole and unscattered" at the cabin.

I have a little different theory about the Hoovers' departure; it has to do with our driveway. The previous owner of our property cut a driveway down to the cabin site right along the Hoover property line.

I have always felt the whining chain saws and snorting bulldozers were the last straw; more than they could tolerate, I'm sure. They left for good shortly after the driveway was completed.

It is true that the neighborhood has changed some since the Hoovers left, but much has remained: the beauty, the wildness, the creatures, the romance of the Boundary Waters. The renewing solitude still reaches deep into the needy souls of those of us who have come more recently to the forest.

There have been many generations of deer around here since the Hoovers came, imprinting forty years of dependency on those gene strings. Corn is an acquired taste for these deer; there isn't a stalk of corn growing within a hundred miles of here. It is their candy, and they come running down the driveway if they see me scattering a bucketful near the woodshed. I have tried substituting the alfalfa-millet pellets the Department of Natural Resources distributes during periods of late winter starvation threats, but the deer ignore them if they think corn might be available.

Certainly these deer are far-removed linear descendants of those that frolicked in Hoover's yard, but occasionally some characteristics Helen described are very apparent. She wrote at length about Mrs. Meany, a large, evil-tempered doe that wouldn't let any other deer near the corn pile when she was feeding. She would rear up on her hind legs and lash out at any trespasser or charge the offender with her head down and kick the fawns as they ran past. For the past couple of years a doe has come into our woods that acts much like Mrs. Meany. The behavior is similar, and she looks exactly like the one Helen described—very long-legged, with an unusual slate-gray color, exotic almond-shaped eyes, and an unusual white blaze on her forehead.

Then there is Pig II, a fat, frowzy little buck fawn that showed up last year; it stepped right out of the pages of *Gift of the Deer.* Helen's Pig was one of twin fawns she named because of his greed and impertinence. The Pig at our corn pile looks and acts exactly the same and is regularly whacked by the other deer as he tries to sneak under the grown-ups to get at the food first. He is always scruffy looking, rejects all discipline, and in every way fits Helen's description of Pig. I suppose it is highly unlikely that behaviors and personality traits travel down the generations, but I choose to think they are all part of an extended family.

Watching the deer in a clearing on a winter morning from the warmth of the cabin, it is easy to recreate this scene forty years ago: No electricity, no telephones, not much of anything in the way of conveniences. Hoovers used the icy lake for refrigeration, and Helen washed clothes in the "Sylvan Glade Laundry," a space cleared near the woodshed for the temperamental gas-operated washing machine. In 1960 Ade cobbled together a pressurized water system from a small engine, hoses, a pump, and an old stock tank. There was an outdoor biffy, of course, but they eventually got a gas-fired indoor toilet that operated on a twelve-volt battery. They were without a car for many years after an accident on the Gunflint Trail totaled their old Chevy. They got their groceries via a once-a-month delivery truck from town.

Out of the corner of my eye I spot two yearlings. They sneak out of the clearing and head for one of the sheltered spots where I sometimes drop a handful of corn. They are very wary and stop every few steps, swiveling their radar-dish ears and pointing their noses into a vagrant breeze . . . cautious, but with great anticipation. The rules for good ecology blur, and I reach for the corn can. Maybe I'll stop feeding them next year.

10
More than a Bird

I TRUDGE SLOWLY UP the wooded hill, moving as quietly and inconspicuously as possible, crouching behind a moss-covered boulder here and a blowdown spruce there. I know the raven (*Corvus corax*) is up there somewhere, probably high in the gaunt and grotesque branches of the dead white pine near the crest of the ridge. It has been "quorking" about something since sunrise. Probably broadcasting the morning's gossip to the rest of the flock. Or perhaps, this morning, it is the watch raven. Ravens have the most sophisticated and varied communications system of all the birds; quorks, mews, whines, yells, gargles, and yips, all with different inflections, are only a few of their vocal tricks. There is general agreement among biologists that ravens are also among the most intelligent of all wild creatures. In *Ravens in Winter,* Bernd Heinrich recalls a fellow professor discouraging one of his graduate students from doing a paper on them because, he warned, "Ravens are smarter than you are."

Ever so slowly, I continued my stalk up the hill, slipping into a shoulder-high patch of thimbleberry bushes near the edge of a small clearing and focusing my field glasses on the top branches of the dead pine. No raven. As I searched surrounding trees, suddenly I heard the vibrating swoosh of those powerful four-foot wings above me as the raven twisted through the branches, moving its head from side to side, its cocoa-colored eyes fixed on me. It seemed to be saying, "Never try to fool me, oaf."

What a bird the raven is! More than a bird. In the Distant Time of the Athabascan family of tribes along the Pacific Northwest coast and Alaska, Raven was the creator of the earth, sun, moon, and stars, stealing light from the people of darkness and establishing human life on earth. In the lives of the Koyukon, Tlingit, Tsimishian, Bella Coola, and other Canadian tribes, the raven was not a supreme being or an object of worship, but a creature of great power and influence in their lives. To them it was also the ultimate trickster, a paradoxical concept that has sometimes puzzled scholars, but never the native people. They communicate with the raven in prayer and invoke its power to combat illness or bad luck. In the confusing (to us) mélange of Christianity and native beliefs that these tribes practice today, the raven is still a very important figure. It sits atop most totem poles, smirking and imperious. Taboos against killing a raven are as strict as those against incest. To the Koyukon people it is Tseek'al, Old Grandfather to everyman.

In spite of its history, its pedigree, and its glory days, in our culture the raven is unknown, unloved, and, what seems to be a worse fate today, unmarketable. Bullwinkle Moose, Bambi Deer, Anyloon, Old Turtle, Smokey Bear, Howland Owl—all these neighbors have been immortalized in books, films, cartoons, product reproductions, and commercials for years, but the raven is passed by. It seems that no one, other than a few Native American artists, pays much attention to it.

"It's because they eat corpses," explains Pat Shunn from high in her bird-watching aerie above Lake Saganaga. She may have something there. The raven's image and reputation could use some buffing up. It is a carrion eater, habitué of garbage dumps, and a legendary death messenger. It has a raucous, in-your-face personality and is not cuddly. Cute, it isn't. But it is brainy, majestic, and fascinating. "A big, black beautiful bird," Heinrich calls it.

Black is its distance color. Up close, in sunlight, the raven is a shimmering mix of dark purple, green, and bronze, with silvery highlights in the long hackle feathers of its cape, which it fluffs out like a bishop's ruff.

Because of their similar color and shape, crows and ravens are often mistaken for each other. They are in the same family (*Corvidae*), but the raven is much larger, with a maturity weight four times that of the crow. It has a four-foot wingspan, and the tail shape is the best field marking in flight. The crow's tail is squared off, while the raven's is rounded, or spade-shaped.

The invisible difference is the raven's vaunted intelligence. However, as Heinrich points out, "It is common to confuse a personal interpretation of intelligence with observation." He emphasizes that there are many ways to see things, and observations differ with observers.

A recent event here, open to interpretation, occurred when Justine Kerfoot and Charlotte Merrick, our two senior coureurs de bois, who poke around in every corner of the forest looking for wildlife, spotted a flock of ravens circling and yelling over a grove of jackpine woods west of the Gunflint Trail. Justine shifted her four-wheeler into low and took off through the woods in the direction of the commotion. Arriving in the general area where they had seen the ravens circling, they got out and began searching around. Near a path they found a dead bear cub, so they crawled under a big white spruce to wait until the ravens came down to feed. But they didn't come down. They continued circling and yelling until Justine and Charlotte lost patience and left. They returned the next morning to find an interesting set of clues. The area was covered with wolf tracks, the remains of the bear, and eight ravens helping themselves.

The conclusion they reached was that the ravens needed a carcass opener and their display was calculated to attract the wolves in order to complete the job they were incapable of. Seemingly a perfect example of how ravens often seek the cooperation of other animals to get a meal. Or was it?

There is a possibility that the ravens were pursuing another of their strategies—following predators to clean up the scraps. The wolves may have been in the vicinity all along but remained hidden while humans were present. Or it might be possible that the ravens were calling in friends and family to share the feast, a common raven practice biologists call recruitment. Whatever the motivation, there was a lively intelligence displayed during this incident.

Ravens seem to have an extrasensory perception, a mysterious sixth sense that goes beyond their superior

eyesight in locating a meal. Ice fishing for lake trout last winter on Lake Saganaga, Frank Shunn had an experience with ravens that highlights this special ability to come out of nowhere and find food. Saganaga is a huge lake, forty miles long, and out in the middle one can see forever. Frank was tending a series of holes he had cut along a favorite reef. When he caught a trout, he would bury it under a mound of snow near the hole and go on to the next one. Before leaving for his next fishing spot, he would look carefully in every direction . . . not a raven in sight. When he got a couple hundred feet away from the buried treasure, he turned quickly and there were five ravens busily digging out the buried fish. Who knows where they came from, or how they knew where the trout was? This remarkable ability to locate food is one of the reasons the raven is the most widely distributed bird on earth. Or might this be another activity of the supreme trickster? In *Make Prayers to the Raven,* Richard K. Nelson says, "Whatever else the ravens may be, they are indeed successful. But then, who should better know how to live on the land than its own creator?" The raven also has a unique ability to make the best of things. If caught in a trap, many wild creatures, notably the mink and the weasel, will chew their own feet off to escape. The raven would perhaps settle down and wait for the trapper to return. Then it would arrange for its release and, most likely, con the trapper out of his lunch.

After a day's trout fishing on Esther Lake off the Arrowhead Trail, Mary Ofjord came over a slight hill and discovered three ravens in the road ahead, jumping around and making an awful fuss. At the sight of her pickup, one of them flew away, one ran into the woods, but the third was badly injured and could hardly move. Mary got her trout net out of the truck, captured the bird, and drove home with it in the net.

The Ofjord home, at Poplar Lake, resembled a mission for the street people of the wild world. There were always strays or infirm and aged, injured and sick creatures in cages, on perches, or maybe just hanging out on the front porch. Mary was their nurse and spiritual adviser and Jon, who had spent several years at the University of Minnesota's Raptor Center in Saint Paul, was licensed federally as a wildlife educator and by the state to take in and treat injured wild creatures.

The injured raven, whom they named Corvus, was a young male whose primary feathers on one wing had been torn out, almost as though it had been pinioned, and the wingtip area was lacerated and badly infected.

They got some amoxicillin from a pharmacy in town, cleaned and bandaged the wing, and put Corvus into an eight-foot-square cage made of framing and chicken wire. They placed this outside, but near the house so they could observe the raven constantly. He liked his new home, especially the food service. Jon and Mary let it be known in the neighborhood that they needed mice, squirrels, groundhogs, or any other road-killed or trapped critters for Corvus. Responding as they usually do to the needs of others, regardless of genus or species, the neighbors came through in grand style. Soon the Ofjord mailbox began to fill with dead mice, some of them scrubbed and in ziptop plastic envelopes. Some people stopped by in their boats and left goodies for Corvus on the dock. Game manager Bill Peterson dropped off a road-killed deer occasionally, and others cruised the Gunflint Trail for squished groundhogs or squirrels. Mary saw to it there was always a good supply of shelled corn, kibbled dog food, and red grapes in Corvus's dish. He made a rapid recovery, even though he was still flightless, and soon became a member of the family. To his credit, he always shared his bounty with others; coyotes, wild ravens, red squirrels, other birds—all were welcome at Corvus's table. He would often tear off a piece of meat and offer it through the cage wire to Muffy, the small black dog of unknown ancestry that had wandered in.

The raven's relationship with Muffy often made Mary wonder if there was any truth to the old adage about birds of a feather flocking together. Coal-colored Muffy was Corvus's special friend, but he would have nothing to do with Scooter, a white poodle that also lived with the Ofjords. Scooter's good buddy was Brewster, a white cockatoo, whose perch was in the basement. They would play together for hours.

The Ofjords had a family of gerbils that multiplied so fast they could not all be kept. Mary, unable to destroy them, would put some of the surplus in Corvus's cage and leave. When she would come back the gerbils would be gone and Corvus would be preening and looking very content. It did answer a question they had wondered about. Is the raven a predator? Will it kill prey? Apparently yes, it does prey on other species. There are stories from the far north about ravens breaking the skulls of caribou calves and young seals with their heavy beaks.

Corvus had another strange habit. He never drank water. When they put a dish of fresh water into his cage, Corvus would jump in and take a bath, but he never drank the water. He did eat snow in the winter, and Jon decided he was getting enough moisture from his food. Jon and Mary had Corvus for eight years, and in that time he became much more than a pet. He was a family member and a great companion to the other animals, except Scooter. One winter night a pine marten tunneled through the snow and chewed a hole in the chicken-wire cage, and that was the end of poor old Corvus. All they found in the morning were some black feathers and the feet.

The years with this interesting bird have had a profound effect on Mary. She still feels the presence of Corvus, and she refers to him as her spirit helper. She is even convinced that someday she will return as a raven.

This is not a unique experience. Many people report a strong spiritual connection with ravens. There is something very different about this bird's ability to communicate and establish a relationship with humans, something charged and very personal. It is a connection with another world that cannot be dismissed. Elijah, the prophet, was saved from starvation in the desert by a God-sent raven. A raven sat on the shoulder of Odin, lord of all Norse gods, as adviser, messenger, and companion, and perhaps as the source of Odin's power. Ravens have occupied similar positions of influence throughout the history of mythology and folklore, and most people, including Edgar Allan Poe, who have had an intimate contact with this remarkable bird have felt its spirit power.

I think about this a lot, especially when the raven comes in the morning and perches in the dead pine in back of the cabin. It doesn't come for food, I'm sure, and it doesn't have a nest nearby. It comes to check up on me, and when I know it's there, I go out and watch it settle itself on a branch. It usually gargles a bit, shifts its weight around, and combs its shaggy neck feathers with its heavy bill. Then it leans forward and peers down at me. I can see kinship in the eyes of a bear, mischief in the glance of a red squirrel, and danger in the golden stare of a wolf, but when the raven tilts its head, looks down at me over that big nose, and spears me with that look, I can feel its power. Determined to bluff it out, I stare back, and if our eyes meet long enough, I'm almost certain I can see the creator of the world and the supreme trickster . . . at least, I'm sure that's what it wants me to think.

11
Icon of the Wilderness

ONCE IN A WHILE ON magic winter nights when the stars seem to prowl the topmost branches of the big pines, I hear the wolves (*Canis lupus*) howl and am transfixed by this wild music—discordant, unreal, hypnotic—that seems a chorus from the spirit world. It is one of the oldest animal sounds on earth, over a million years, and I thrill to it now out on a frozen lake in northern Minnesota, watching satellites ranging over electronic trails across a midnight sky.

"Nature's pipe organ," someone called the music. It has an intense, complex power. Some imagine shadows drifting through the forest and are fearful. To the biologist it is a natural sound bite, encoded but translatable. To others it is a musical experience, and to the naturalist it is the purest note of wilderness freedom—what Durward Allen called "the jubilation of the wolves . . . singing with glee club exuberance." Two residents along this shore of Gunflint Lake enjoy the wolf concerts so much they have devised a way of recording them by placing microphones outside, wiring them to their stereo sets, and pumping the music into their living rooms.

Sometimes the wolves may be on a high ridge in back of the cabin, or they may howl from the Canadian side of the lake and a response will come from somewhere far to the east in the headlands. It is always difficult to tell how many wolves are howling. They change pitch often and wail in a bluesy counterpoint, and sometimes a solo tenor will take off high above the others, but all this may be only two wolves.

L. David Mech, one of the nation's preeminent wolf researchers, listens to wolf howls with a scientist's ear, searching for meaning and messages. His studies indicate that wolves howl for a variety of reasons, but primarily to inform other wolves about the size and location of the pack, to identify individual wolves, and to broadcast territorial information.

However, to the lay wanderer of the winter roads, when the howls well out from the frozen forest or carom off the granite escarpment above the river, it is difficult to think of wolves as "solidifying social bonds." Even though I know they aren't going to devour me and I always thrill to the music, my breath stops momentarily and the cabin seems very far away. The portly retriever who walks with me feels the same way as she quits her exploration of the ditches and presses tightly against my leg.

The fear and apprehension that many feel when hearing the wolves howl is a psychic echo of a two-thousand-year-old tape playing out old myths and legends that portray wolves as brutal and indiscriminate killers attacking travelers on moonless nights and digging corpses out of fresh graves. In this folklore of the evil of wolves and their habitat in remote and wild places, they are "creatures of waste and desolation," and both wolves and wilderness should be destroyed.

In *Of Wolves and Men,* Barry Lopez insists that fear and righteousness are what killing wolves is all about: "The hatred has religious roots. The wolf is the devil in disguise . . . and man demonstrated his own prodigious strength as well as his allegiance to God by killing wolves and subjugating wilderness." This is a distinctly European attitude. With the exception of the Navajo, who believed that the wolf was a witch in animal's clothing, most Native Americans perceived the wolf as a spiritual brother. The Ojibwe of this region, who had lived with the wolves for centuries before the white man came, saw Myeengun, the wolf, as a medicine animal, a totemic symbol of guardianship and perseverance. They thought it was the best parent among the animals and admired its hunting ability—a thoroughly exemplary citizen in the Ojibwe view.

Its killing skills and gross eating habits have always been the source of the wolf's public relations problem. Both the loon, that wilderness darling of millions, and the wolf are carnivores and predators, and both obey the first law of their kingdom: kill to survive. The difference is in the public perception. The wolf kills Bambi deer and bunny rabbits out in the open, while the crafty loon kills salamanders, fish, leeches, crabs, and other scuzzy creatures under water and out of sight.

The white-tailed deer is the main item on the wolf's menu here, with an occasional moose, beaver, rabbit, ground-nesting bird, or even domestic dog or cat, if it wanders too far from home. A wolf pack will gang up on a moose, but a single wolf will seldom take that chance. The normally docile giant is a very dangerous opponent, and its flailing hooves can pulverize a wolf.

In this boreal forest, nature can always be counted on to conceal her most dramatic moments. Bear cubs are born to feed and frolic first in the hibernation den. The elegant loon, drifting sedately on the lake, becomes a dagger-shaped missile beneath the surface; the flying squirrel glides in silent and daring flight in the dark of night. Only occasionally will we witness one of those sudden and vivid episodes that make all the waiting and watching worthwhile.

One bright February morning Julie and I were enjoying our steaming mugs of coffee, the noisy warmth of the wood stove, and the view across the lake over a mile of wind-etched snow into the brooding spruce forest of the Ontario hills. Since November, when the deer returned from their annual summering in the eastern hills, nearly sixty inches of snow had fallen, most of it driven by the northwest wind that had sculpted long graceful folds hanging from the eaves and intricate abstractions in the tangled screen of alders along the shore. From the big window it was the quintessential peace of winter morning in the Boundary Waters.

This tranquillity turned savage in an instant as a terrified deer, having committed the fatal error of allowing itself to be driven out onto the lake by wolves, appeared in full flight a short distance from shore. It was pursued by a large, brindled wolf who seemed to be running almost casually. The wolf's heavily furred paws kept it on top of the crusty snow, which the deer's pointed hooves penetrated with each frantic bound. Having examined hundreds of blood-stained kill sites and read many accounts of wolf kills, the inevitable attack was not at all what I expected. It was suffocation rather than a bloody, fang-and-claw killing.

Running almost side by side with the deer, the wolf leaped high and bit deep into the neck, twisting its body in the air as speed and gravity plowed both the doomed deer and the wolf into the snow. The wolf's jaws were clamped over the larynx, and it hung on as the buck dragged it in an irregular circle, trying desperately to regain footing. This struggle lasted only a brief minute before the deer collapsed, front legs pawing futilely at the wolf's deadly strength, back legs jerking spasmodically. The wolf, almost motionless, kept its jaws locked over the deer's throat until the struggling ceased. There was no ripping and tearing; the powerful jaws clamped over the windpipe and about eighty pounds of dead weight were the wolf's only weapons.

The bloodletting began shortly after the kill, when the executioner was joined by two other wolves of identical size and pelage—one from across the lake and the other from the west, indicating that an ambush had been set, a common hunting strategy of the wolves. The three began feeding immediately, and keeping close watch on the time, we were surprised to see that the deer was almost entirely consumed in just over forty-five minutes. Well fed, with bellies dragging

in the snow, the wolves waddled off, one of them dragging a haunch, to a sunny bluff on the Canadian shore, where they all napped for several hours.

We were anxious to examine the kill site more closely and hustled out on the crusty snow of the lake in front of the cabin. A few pieces of bone, some ragged tatters of hide, the tongueless skull, and the untouched rumen (the carnivorous wolf seldom touches the stomach contents of an herbivore) were scattered over a blood-stained patch of deer hair approximately four hundred feet square. I have often wondered if the deer shares with the domestic cat the ability to release great amounts of hair during moments of extreme stress. There appears to be no other way to explain the deluge of deer hair that covers a kill site. Before we arrived, a dozen ravens, nature's super-efficient janitors, were on the job, cleaning up. As we poked around, they scolded and backed off indignantly, but kept sneaking in for a choice morsel when our backs were turned. We took some photographs, marveled at the deep tooth grooves in bone fragments, and measured tracks, which were five and one-half inches wide, normal for an adult timber wolf.

Back in the cabin, we watched with field glasses for an hour as the wolves snoozed, rolled around in the snow, and frolicked like puppies on the sunny shore across the lake. The ravens took everything they could get away with, and by the next morning nothing was visible except a shadow of deer hair constantly shifting and changing shape on the blowing snow.

These wildest of our neighbors are one of many small packs that make up the only viable wolf population in the lower forty-eight states. They are somewhat protected from hunting, trapping, and poisoning, but face an even more serious threat here from the destruction of their habitat by logging, road building, and development. So now, when I see those great, tufted tracks in the snow circling an otter's airhole in the ice, or hear the wild music cascading down the ridge on winter nights, I can't help but wonder how much longer they'll be here. I'm afraid it won't be too many years before they'll join the wolverine, the woodland caribou, and the other wilderness ghosts.

12
Ghosts

THE OLD-GROWTH forest is spirit country: deep shade, strange night noises, distorted faces in the foam at the edges of rapids, spirit fires smoldering along ancient hunting trails, footfall sounds crossing the cabin roof at night that disturb sleep but leave no tracks. Ghosts of all kinds. Phantom canoes full of singing voyageurs have been seen down among the islands near the long portage. Sometimes I find that the weird night moaning is only trees rubbing together in the wind, but mostly I believe in the ghosts.

Because this is such an ancient habitat, predating history, it is certain that many of the ghosts are animals. There was a time when this region was home to woolly mammoths, saber-toothed tigers, giant sloths, and beavers the size of cows. Those ancient creatures disappeared as nature tinkered with design and adaptability, carrying on the imperceptible work of evolution. Only the beaver has left a traceable lineage. All the rest have wandered the long trail to oblivion.

Some, more recent residents, have been the victims of our tampering with the environment, and the result has been extirpation, a word implying removal. It doesn't have the finality of "extinction," which means "no longer in existence." We have quite a few animals that fit that extirpation pattern. They aren't really gone, but are imagined, or seen so seldom, that theirs is a ghostly presence.

Some, like the fisher, seem reluctant to leave and occasionally show up in trapper's sets, or are sometimes seen as a flash in the undergrowth or hiding in a clump of witches'-broom, the wildly growing mass of twigs and foliage that forms in spruce and balsam trees. Others, like the cougar, or mountain lion, seem an anomaly in the boreal forest. They are all spectral creatures to me. I have never seen any of them, but I know they are here. Occasionally a set of tracks is seen, a corpse is found, a sighting is well documented, or, if we are lucky, a photograph is made. That's what happened with wilderness guide and writer Harry Drabik's woodland caribou.

Woodland Caribou (*Rangifer caribou*)

The barren ground caribou of the far north is moving constantly, migrating thousands of miles every year, but the woodland caribou is more of a wanderer than a migrator. There are still some caribou in Ontario, next door to us, and one year some of them came south to Minnesota for the winter. Drabik told his story to the *Minnesota Volunteer:* "I wasn't expecting anything, but as our vehicle rounded a turn in the road, my companion and I saw what looked like a moose standing about 200 yards ahead. There was something odd about that moose. My companion said it for me. 'It's sure got a funny rack for a moose.' Just then the animal turned away and both of us caught a white flash on the animal's rump."

What they saw was a woodland caribou, but Harry was reluctant to talk about it. "It's sort of like reporting a UFO," he said; "most people think you are either loaded or goofy."

Eventually his curiosity got the best of him, and he reported it to the DNR area wildlife manager Bill Peterson in Grand Marais, who was a little reluctant to accept the sighting until David Mech spotted the caribou from the air a short time later. Peterson knew the procedures necessary for official verification, so he had Harry guide him to the location, where they collected scat and hair samples. After trailing the animal for several hours, they got a good set of photographs. It was a woodland caribou, all right, the first one seen in Minnesota in fifty years.

I remember the excitement that Harry Drabik's caribou caused here. There were a few old-timers around who still remembered when these woods were home to the caribou, and now there was speculation that they were returning. I went back to my journal looking for an item I vaguely remembered about a plan by the Safari Club to reintroduce the woodland caribou into this region many years ago. I finally found it in a 1977 entry: "Biologists were brought down from Canada and pronounced the BWCA [Boundary Waters Canoe Area] a viable habitat. The project would involve two herds, one at the Minnesota Zoo and one penned somewhere in the BWCA. The latter would be pregnant females and calves. The young born to this herd would not have the imprint of a home territory, so would hopefully stay put. . . . cost is the big problem. . . . project is on hold."

Indeed cost was a big problem, so big the project was eventually abandoned. But now, twenty years later, the project is being revived. If it works we may see the resurrection of a ghost, but I'm not going to hold my breath. I have seen computer models crash and burn here in the bush country, but I'm sure the wolves will think it's a great idea.

Wolverine (*Gulo luscus*)

"Devil Beast" is what the Indians called the wolverine, the largest member of the mustelid family, cousin of the weasel, the fisher, and the badger. About forty pounds of strength and fury, the wolverine regularly challenges bears and mountain lions for prey. Wolverines have been officially extinct here since 1940, but about a dozen have been killed by hunters in recent years, and tracks were identified last winter here in this neighborhood. Not long ago, personnel from the Ontario Lands and Forests Department videotaped a full-grown wolverine thirteen miles from the end of the Gunflint Trail.

Even at its population peak here, it was never numerous, and the range data in *The Encyclopedia of Mammals* lists the wolverine as a creature of the Arctic, subarctic, and taiga. The maps don't indicate a range within three hundred miles of here. But the wolverines are here . . . sometimes.

It is an animal of very remote areas, and that's probably a good thing. It has an awful reputation—greedy, smelly, and notorious for its gluttony. Anders Bjarvall and Audrey Magoun, in *The Encyclopedia of Mammals,* quote this description of its gross behavior:

> It is an animal which feeds on carcasses and is highly ravenous. It eats until the stomach is as tight as a

> drum skin, then squeezes itself through a narrow passage between two trees. This empties the stomach of its contents and the wolverine continues to eat until the carcass is consumed. Although it is a carrion eater, it is also a ferocious predator and will take on anything. It drags down larger prey, such as caribou, by jumping on its back and hanging on with powerful claws until the animal collapses.

It is often mistaken for a young bear and, in fact, was considered a member of the bear family until 1910. It is a valuable fur bearer; there is a city in Siberia whose economy is dependent on the sale of wolverine hides. So valuable is it considered there that the city flag features a running wolverine. In the far north the fur is used primarily for parka ruffs because moisture from breath doesn't freeze on it, almost as though each hair contained a few drops of antifreeze.

From all reports, the wolverine would not make a very desirable neighbor, so I am content with its ghostly status, but it is one of the wilderness's children and its occasional presence adds to the ambience here.

Fisher (*Martes pennanti*)

A few years ago, Duncan McDonald, who has a summer cabin nearby, opened it in the spring and found the interior a shambles—furniture wrecked, rugs ripped up, clothing and other items scattered all over. It looked like a case of mindless vandalism, but when Duncan looked closer he discovered a dead fisher in his favorite armchair. As he reconstructed the crime, it became obvious what had happened. The fisher had come down the tin chimney pipe into the woodstove, opened the door, and jumped into the living room. It found a few goodies in the kitchen, but when it tried to leave, it couldn't get out. The fisher couldn't crawl back up the slippery inside of the chimney pipe, and there was no other exit. In its panic, it tore up everything and eventually collapsed of starvation in the armchair.

The fisher looks like a pine marten on steroids, very similar in shape and color, but much larger. A full-grown male can weigh up to twenty pounds, but the average is closer to fifteen. It is big enough to kill a fawn, but its regular diet consists of mice, voles, snowshoe hares, red squirrels, and deer carrion, mostly roadkills.

There is a myth about the fisher's method of killing porcupines, that it can flip a porky over to get at its belly without getting a snootfull of quills. Only half true. It is at about eye level with a porcupine, and it attacks head on, biting at the facial area. The fisher is lightning fast so it can strike and retreat, strike and

retreat, until the porcupine is weakened by the attacks, and then the fisher flips it over and goes after the quill-less ventral surface.

Bill Berg, furbearer specialist and research biologist, reports that the fisher population reached its peak here in 1960 and has been declining ever since. Because it is the most valuable of furs taken here, it is still trapped, but that is not the reason for its disappearance. Our fisher is doomed because it is a creature of the midsuccession forest, and our woods are getting far too elderly to support a viable population now.

Cougar (*Felis concolor*)

At first people were reluctant to talk about mountain lions being seen around here. There must be some mistake, they said. Like alligators in Arizona. Everyone knows mountain lions are animals of the western landscape—the high passes, boulder-choked canyons, and ponderosa-edged mountain meadows. Elk country. Big sky country. But not here. Not in the lakes and pines country of the north woods.

But they are here. We don't know the numbers, and won't until the cats are radio collared. The mountain lion, or cougar, has such a big range that the one perched on the pile of planks in a Duluth lumberyard and the one crouched at the side of the Gunflint Trail could be the same animal. One hundred miles is nothing to a cougar.

Why not cougars here? The big cat is fairly adaptable as long as it has food, cover, and space, lots of space—all commodities this region has in abundance. Then, too, the Quetico Provincial Park across the Canadian border has another two million acres of wilderness.

Last year there were fifty verified sightings of cougars in northern Minnesota. These weren't the fuzzy, maybe-yes, maybe-no kinds of sightings. The lumberyard cougar sat in front of a crowd of a hundred people, including police and media, with everyone taking pictures and admiring the tawny coat and the long, swishing tail. The mating pair at the edge of a field near Floodwood was observed by two game biologists. And there was the one following a road hunter, while his wife drove the car following the cat. These were cougars, or mountain lions or puma, or panthers, as they are called in Florida.

One night our neighbor Sharlene was driving up the Gunflint Trail a few miles from our cabin when her headlights picked up two glowing coals of green fire. A cougar . . . right near us. It crouched at the side of the road motionless, except for the great, ropy tail weaving snakelike over the gravel. Sharlene stopped and stared at the cougar . . . and the cougar stared back. Less than one hundred feet separated them. The big cat flinched first, and, rather than jumping away, it seemed to just melt into the forest. The cougar is a paragon of ghosts—pale, silent, menacing, incorporeal; not even a dead one has ever been found near here. Until I see one, there will always be a tiny, lingering doubt, in spite of all the evidence.

Windigo

Once in a while, for no apparent reason, a canoe paddle breaks suddenly in the middle of the lake, a compass needle spins crazily deep in the forest, a fire suddenly blazes up in the midle of a swamp. What's happening here? It is most likely the Windigo, the Evil One, a neighbor I don't want to meet, playing his dangerous tricks in this wild country.

Spelled Weendigo, Wiindegoo, or Windigo, this malignant spirit is a centuries-old presence from Ojibwe and Cree legend retold over generations, from Lake Superior to far northern Ontario. According to this legend, there was once a man named Weendigo who lived on the north shore of Lake Nipissing in Ontario, where people were dying of starvation. Driven by the madness of hunger, Weendigo began hallucinating people as beavers, and he committed the unspeakable act of cannibalism. In *Ojibway Heritage,* Basil Johnston tells the story: "Weendigo's hunger superseded everything else. He needed first to be served and satisfied. In James Bay, Weendigo ate and ate, killed and killed." As a result, Weendigo grew to giant proportions and terrorized the land. Eventually he was killed by Megis, who had enlisted the medicine man's power to enable him to challenge the marauder. Unfortunately, death was not the end of Weendigo; he continued on as an incorporeal being, signifying the evils of excess and fiendishness. Even today there are those who attribute inexplicable occurrences to the Windigo. An accident, a freak storm, or any powerful incident that seemingly has no cause is blamed on the Windigo.

One day on the shore of Mine Lake, a few miles out on the Kekekabec Trail, a neighbor who is the soul of sobriety and rectitude watched a phenomenon that left him shaken. It was a hot, still August afternoon, and he was taking a break from his hike, sitting on a big rock at the lake's edge. Suddenly he noticed waves beginning to rise quickly on the surface of the lake. In a few moments there was real turmoil. Waves stood on end and fell back on themselves. He wasn't terribly alarmed, just very curious, until he suddenly realized there was no wind. The trees along shore were perfectly still. He watched awestruck, for several minutes, and then quite suddenly the lake was quiet again.

Later, reporting this incident to a Cree friend living nearby, he got the terse explanation: "Windigo."

La Blaine

This palpable awareness of ghosts in the wild neighborhood is made more manifest by a visit to our ghost town across the lake. For some unknown reason, it was named La Blaine, and, according to regional historian Willis Raff, it had a population of about five hundred in 1893. Even more incompatible with its wilderness setting, it was a railroad town, the field headquarters of a makeshift railroad built to carry iron ore from the Paulson Mine, at the west end of Gunflint Lake, to Port Arthur, Ontario (now Thunder Bay). It was the Port Arthur, Duluth, and Western Railroad, cynically referred to by residents as the Poverty, Anguish, Despair, and Want, or just the Pee Dee. It never actually carried a load of iron ore, or anything else, and eventually went belly up, as did the mine. Shortly after, the president, who was also city treasurer of Minneapolis, went to jail for embezzlement and fraud. The tracks were torn up, and the route along the Canadian side of Gunflint Lake is now a beautiful hiking and cross-country ski trail. La

Blaine died, too, and the forest reclaimed the site. All that remains are some rotting logs and three crumbling, moss-covered ovens—dome-shaped rock structures built by the Italian workers who had been imported to build the railroad and in which they baked their national bread.

A legend persists that there were still engines on the tracks when the railroad folded, and, rather than return them, the crews ran them off the tracks. Nothing has ever been found, but the storytellers say the engines have probably sunk to the bottom of the bogs by now. So, if we stretch our imaginations a little, we can add two railroad engines to our ghost list.

La Blaine was very real, though, a regular boom town. In addition to the railroad barns, there were a hotel, a company store, some cabins, a barracks-like building, and a few saloons. It must have been a lively place then, but only the big pines remember now. The town, the railroad, and the mine all are gone, and that brings up several interesting speculations on the history of entrepreneurs failing to bend this rugged land to their economic will.

In *Pioneers in the Wilderness,* historian Raff tells about the fever of excitement over the glowing prospects of the prosperity that mining of iron and other minerals was sure to bring to Cook County as the century turned. A. De Lacey Wood, flamboyant editor of the *Grand Marais Pioneer,* often got carried away by his optimism. In 1893 he wrote in an editorial: "We have no misgivings as to the future, as the inexhaustible resources of this great region, now soon to be opened up, guarantee an era of prosperity for years to come . . . a region that cannot be surpassed in mineral wealth looming up in every direction, only waiting for the capitalist and miner to develop."

Wood envisioned Grand Marais as the "Pittsburgh of the North." Iron, silver, nickel, copper, and gold were the minerals that would drive the industrial engines for years and make Grand Marais the giant among Great Lakes ports.

It didn't work out that way. Very little in the form of marketable resources was ever found, and all the dreams faded.

Then came logging, which left the county and the exposed spine of the world, the Canadian Shield, a smoking, barren wasteland. After a while an agricultural economy was attempted. There are pictures in the historical files of waving barley fields and pastures of dairy cattle on the slopes of the now heavily forested Sawtooth Range. Potatoes, however, were the big cash crop, and the irrepressible De Lacey Wood cranked up his optimism once more in the *Pioneer:*

"So what if we have, or have not, iron here. It makes no difference. . . . Cook County raises the finest potatoes in Minnesota. In these times a good supply of potatoes beats an iron mine."

Longtime newspaper publisher Ade Toftey remembered when his brother, who sensed a potato boom, built a large warehouse on the site of a present resort complex in the town of Tofte and bought up all the potatoes he could locate, waiting for the price to go up. Unfortunately, he neglected to insulate the warehouse and all the potatoes froze solid, an ironic metaphor for most of the soaring schemes to wrest prosperity from this harsh land.

One day, poking around in the woods at Oven Bay near the ghost town, looking for artifacts with a neighbor who has lived here half a century, we were talking about all the visions of prosperity that have gone awry in this grand old forest. He was thoughtful for a while, seated on a moldering cabin base log, looking out over the bay to an island that once held a transfer warehouse for beaver pelts during the days of the fur trade. Then he looked at me with twinkly old eyes and said, "Maybe God is telling us to leave it alone."

Recommended Reading

History

Blackwell, Billy, and the Grand Portage Local Curriculum Committee. *A History of Kitchi Onigaming: Grand Portage and Its People.* Cass Lake, Minn.: Minnesota Chippewa Tribe, 1983.

A comprehensive presentation of the Grand Portage Reservation history of and by the native Ojibwe people.

Drache, Hiram. *Taming the Wilderness.* Danville, Ill.: Interstate, 1992.

One of the seven books of northern Minnesota history that Hiram Drache has written. It is a chronicle of the wilderness hardships, determination, and good humor of northern Minnesota residents between 1910 and 1939.

Gilman, Carolyn. *The Grand Portage Story.* St. Paul: Minnesota Historical Society Press, 1992.

An exciting and accurate history of the people, geography, and events at the crossroads of the American fur trade, from 1730 to the present.

Johnston, Basil. *Ojibway Heritage.* Toronto: McClellan and Stewart, 1976; Lincoln: University of Nebraska Press, 1990.

The Library Journal says, "This compilation of the basic beliefs, values and ceremonies of a large North American native group is a beautiful and sacred document."

Kohl, J. G. *Kitchi-Gami: Wanderings around Lake Superior.* London: Chapman and Hall, 1860. Reprint, Minneapolis: Ross and Haines, 1956.

Kohl, a German ethnologist, roamed the Great Lakes region in the mid–nineteenth century on foot, on horseback, and by canoe, living with the Ojibwe, whom he found "strikingly like my own people."

Mackenzie, Alexander. *The Journals and Letters of Sir Alexander Mackenzie.* Edited by Wayne K. Lamb. Cambridge: Cambridge University Press, 1970.

The diaries and correspondence of a fur trader, explorer, and entrepreneur, the first man to traverse the continent by canoe from the Atlantic to the Pacific.

Raff, Willis H. *Pioneers in the Wilderness.* Grand Marais, Minn.: Cook County Historical Society, 1981.

A detailed and readable history of Cook County, Minnesota—the explorations, mining, logging, dreams, schemes, politics, and personalities of the North Shore, the Gunflint region, and Grand Marais—by its official historian.

Smith, James K. *David Thompson.* Toronto: Fitzhenry and Whiteside, 1975.

A young reader's biography of a wilderness adventurer who spent his life, from adolescence to old age, traversing the canoe routes of the fur trade from New England to the Rocky Mountains.

White, J. Wesley. *Historical Sketches of the Quetico-Superior Area.* Duluth: U.S. Forest Service, 1967–70.

An informal history of the Quetico-Superior region in folio form, by a forester who spent most of his professional career in the Superior National Forest.

Woodcock, George. *The Hudson's Bay Company.* New York: Crowell, Collier, and Macmillan, 1926.

One of several histories of the largest and oldest fur trading company. Founded as the Governor and Company of Adventurers Trading into Hudson's Bay in 1670, the company is still in the retail business in Canada.

Wildlife

Allen, Durward. *Wolves of the Minong.* Boston: Houghton Mifflin, 1979.

An easily read scientific study of the Isle Royale wolf population by one of the first and most prolific wolf researchers.

Bent, Arthur Cleveland. *Life Histories of North American Jays, Crows, and Titmice.* 1946. Reprint, New York: Dover, 1964.

An older but still serviceable ornithology text, well written and prized by birders. Written by one of North America's premier ornithologists.

Bolz, Arnold. *Portage into the Past.* Minneapolis: University of Minnesota Press, 1967.

Bolz, his wife, and a Crane Lake canoe guide follow the Voyageur's Highway from Grand Portage to Rainy Lake, reading the diaries and journals of early travelers as they go.

Casey, Denise. *The American Marten.* New York: Dodd, Mead, 1988.

A well-illustrated young reader's life history of the American pine marten.

Eckert, Allan W. *The Owls of North America.* New York: Crown, Weathervane Books, 1987.

The definitive encyclopedia of North American owls. Authoritative but entertaining, with a lot of interesting peripheral information about regional names and cultural history.

Fair, Jeff. *The Great American Bear.* Minocqua, Wis.: NorthWord Press, 1990.

Written with the assistance and photography of Lynn Rogers, this book weaves together the daily lives, cultural significance, and management science of Minnesota's black bears.

Goodchild, Peter, ed. *Raven Tales.* Chicago: Chicago Review Press, 1991.

A collection of worldwide raven folklore and mythology.

Hazard, Evan B. *The Mammals of Minnesota.* Minneapolis: University of Minnesota Press for the James Ford Bell Museum, 1982.

This is the essential habitat guide of the 81 mammal species of Minnesota. The complete glossary is very helpful if the reader is confused by biological terms.

Heinrich, Bernd. *Ravens in Winter.* New York: Summit Books, 1989.

Heinrich, a professor of zoology at the University of Vermont, gives a remarkable account of his years living with ravens at his Maine camp.

Holmgren, Virginia. *Owls in Folklore and Natural History.* Santa Barbara, Calif.: Capra Press, 1988.

The only gathering of owl lore, mythology, and legends

combined with solid natural history. If owls beguile you, this is a must read.

Hoover, Helen. *The Gift of the Deer.* New York: Knopf, 1973.

The award-winning, best-selling account of the Hoovers' life with their neighbors, the deer, at their Gunflint Lake cabin fifty years ago.

Janssen, Robert B. *Birds in Minnesota.* Minneapolis: University of Minnesota Press for the James Ford Bell Museum, 1987.

A field guide to the distribution of four hundred species of birds in Minnesota, complete with range maps for each species.

Kerfoot, Justine. *Gunflint.* Duluth, Minn.: Pfeiffer-Hamilton, 1991.

An engaging journal account of a long life in Minnesota's Gunflint region, with a delightful cast of characters: guides, Indians, hunters, fishermen, animals, and early tourists. A charming biography of the region and its residents by Our Lady of the Trail.

Klein, Tom. *Loon Magic.* Ashland, Wis.: Paper Birch Press, 1985.

A loving celebration of the loon—its habitat, distribution, family, migrations—and its future, with a preface by Sigurd Olson. Superbly illustrated with color photographs.

Longley, William, and John Moyle. *The Beaver in Minnesota.* Minnesota Department of Natural Resources, Division of Game and Fish, Technical Bulletin no. 6. St. Paul, 1963.

The label "technical bulletin" is a little misleading. This is a complete profile of the Minnesota beaver population's history, biology, distribution, habitat, and management, written for lay persons.

Lopez, Barry Holston. *Of Wolves and Men.* New York: Scribner, 1978.

In this wolf classic, we see a variety of wolves, from the view of the Native American, the scientist, and our own imagination. John Fowles says, "A remarkable book, both biologically absorbing and humanly rich."

Macdonald, David W. *The Encyclopedia of Mammals.* New York: Facts on File, 1984.

Everything you ever wanted to know about mammals. A well-illustrated and complete desk reference.

Matthiessen, Peter. *Wildlife in America.* New York: Viking Press, 1959.

This is a comprehensive history of American wildlife that focuses on threatened and vanishing species. Written by one of our great literary naturalists.

McIntyre, Judith W. *The Common Loon: Spirit of Northern Lakes.* Minneapolis: University of Minnesota Press, 1988.

The book jacket describes this as a "lively, authoritative account of one of our most primitive and spectacular birds." It is beautifully written by a professional ornithologist, who reminds us of the many environmental threats the loon faces.

Mech, L. David. *The Way of the Wolf.* Stillwater, Minn.: Voyageur Press, 1991.

The definitive work on Minnesota wolf populations, by the country's leading wolf researcher.

Nelson, Richard K. *Make Prayers to the Raven.* Chicago: University of Chicago Press, 1983.

This is an ethnographic study, but that should not intimidate the reader because it is also a great adventure story, a meticulous description of the Yukon and its people, a record of raven legends, and a sensitive examination of a disappearing culture.

Roberts, Thomas Sadler. *The Birds of Minnesota*. Vol. 2. Minneapolis: University of Minnesota Press, 1937.

The classic volume of Minnesota birds, with watercolor plates by Walter J. Breckenridge, Frances Lee Jacques, and Louis Agassiz Fuentes. Out of print but available at some schools and libraries.

Rue, Leonard Lee, with William Owen. *Meet the Beaver.* New York: Dodd, Mead, 1986.

An account for young readers of a first beaver experience, which invites further study. Complete, accurate, and fascinating reading.

Shepard, Paul, and Barry Sanders. *The Sacred Paw.* New York: Viking Penguin, 1955.

A detailed and delightful look at our long fascination with bears.

Terres, John K. *The Audubon Encyclopedia of North American Birds*. New York: Knopf, 1980.

The basic and definitive bird book, beautifully illustrated with color photographs of the species in their natural habitats.

Van Wormer, Joe. *The World of Moose*. Philadelphia: Lippincott, 1972.

A comprehensive biography, biology, and adventure that concentrates on the Alaskan moose, by one of the country's foremost wildlife writers.

John Henricksson is a writer and editor who divides his time between Mahtomedi, Minnesota, and his Gunflint Lake cabin. He is the author of *Rachel Carson: The Environmental Movement* and editor of *North Writers: A Strong Woods Collection* (Minnesota, 1991) and *North Writers II: Our Place in the Woods* (Minnesota, 1997).

Betsy Bowen is author and illustrator of *Antler, Bear, Canoe: A Northwoods Alphabet Year, Tracks in the Wild,* and *Gathering: A Northwoods Counting Book.* She lives in Grand Marais, Minnesota.